Design

Ecology

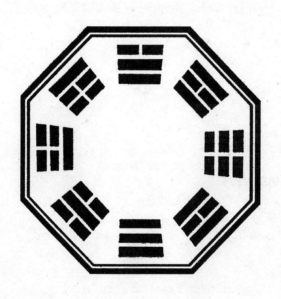

MICHAEL RIVERSONG

ISBN   0-945685-17-3

Library of Congress Catalog Number: 94-071208

10   9   8   7   6   5   4   3   2   1

To contact the Author:
    Quality Environmental Assessments
    P.O. Box 2775
    Cheyenne, Wyoming 82003
    (307)635-0900
    (303)829-0774

Published by:
    Borderland Sciences Research Foundation
    P.O. Box 220, Bayside, CA 95524

THANKS

The following people were instrumental in the process of getting this book together. This is a thanks to them:

Steve Elswick, of the International Tesla Society and Extraordinary Science Magazine, who convinced me that I really did know something worth sharing.

Helmut Ziehe, the Grand Wizard of Bau-Biologie, who has done more than anyone else in the Western Hemisphere to get ecological living principles out to people who need it.

Dr. Paul Finney, for referring me to enough clients all over central Kansas to be a huge factor in building the experience which went into this book.

Wolfgang Maes, Germany's foremost Bau-Biologist, an inspiration to all the pioneers in this field and a great interviewee.

Roberta Rivers, my former wife, who put in a lot of time, energy, money, faith, and love during the initial phases of this book, before I knew it would become a book.

Carol Wagner and Karen Mohr, at the International Association for New Science, two heroes who helped set up an essential forum for these ideas.

Joseph McFadden, a great Chemistry teacher, who doesn't let anyone get away with anything. We need more teachers like this.

Louise Aron, my attorney, who helps simplify and clarify things.

Michael Thau, a friend who deeply believes in improving life on this planet, and devotes his whole life to that. His assistance in printing the first edition was essential.

Michael Theroux, Director of Borderland Sciences Research Foundation, who saw this book through to publication and made sure it happened.

INTRODUCTION

Since the first edition was printed in May, 1993, this book has already made differences in my life and the lives of the first wave of readers. Letters and phone calls slowly trickle in, and sales have continued even without any promotion at all. This has been encouraging, and has led to the publication of this third edition, which you are now reading.

WHY DESIGN ECOLOGY?

Lots of people are having problems with their living and working environments. Unfortunately, some people are creating fear about these problems, trying to carve out a niche for themselves and make some money on experimental remedies. This is really unnecessary. We will all go a lot farther and be capable of more honesty if we sell expertise in solutions. A good consultant should be a detective and educator, and let others sell remedies. This encourages objectivity. After all, the fears are there. So are the problems. If we concentrate on solutions, then this discipline will mean something over a long term. So a responsible practice of Design Ecology involves fearlessly focusing on the solutions to problems. As we do that, everyone gets to move up in health and happiness.

INTRODUCTION TO THE FIRST EDITION

We can go on and on about how the planet is getting destroyed, and our species probably won't survive, and all that. Some people are convinced there's no hope, and are in the process of partying until they're broke. It is possible that Nature will accomplish the cleanup process for us, without our help. If that happens, our species actually will probably survive, but in greatly reduced numbers, in a sort of New Dark Age. Or we can look carefully at the mess we've created, and figure out how to clean it up.

TABLE OF CONTENTS

PART I

FOUNDATION OF THE SCIENCE

OVERVIEW

Humans are obviously designed to take control of things. We have incredibly flexible hands, relatively good mobility, decent perception of what's near us, and mind powers which we are just beginning to find. Notice that when someone feels like there is no possibility of control, you have a depressed or apathetic person on your hands. The more in control you are, the better you tend to feel. This does not mean control of people. Our evolution, both culturally and physically, has been consistently in the direction of greater control over environmental factors.

Control over the environment does not necessarily mean bulldozing the nearest hill. It also doesn't mean setting up government bureaucracies to micro-manage every aspect of construction, manufacturing, and trade. Actions taken without due consideration make the environment more unpredictable. For optimum control, intelligence needs to be applied. Good design is the environmental manifestation of applied intelligence. Actually, even small amounts of control exerted over an environment will help. If things are overwhelming, as in dirty laundry piled up everywhere, walls coming apart, ceilings falling down, you can start by emptying the stupid ashtrays, and you'll start to feel better. That bit of control can lead to more and more control, until eventually many key problems are solved, and life can get better.

Once an environment is fully suitable for humans, our potential increases dramatically. Most people feel a sense of spiritual bondage, and many religious practices have evolved to help us at least have some hope of being released from this state. However, if an environment is unsuitable and becomes too much of a problem, it may be impossible to pursue anything which would truly and permanently enhance spiritual growth. Therefore, environmental design has a spiritual dimension, and is a way to gently move us along to a better state of being.

During the past few years awareness of alternative design techniques such as Feng Shui and Bau-Biologie has been steadily increasing in America. These techniques offer the individual a chance to control environments, without government intervention. Professional Feng Shui practitioners are known to be thriving in Los Angeles, Berkeley, Las Vegas, Denver, New

York, Dallas, and Boston. This makes Feng Shui into a growing and exciting field of consulting. Meanwhile, The International Institute for Bau-Biologie and Ecology is now operating in Clearwater, Florida, certifying practitioners through seminars and a correspondence course. Several other organizations scattered around America are working with concepts closely related to these two branches of science.

What all of these emerging groups have in common is a concern with design principles as related to the total environment experienced by humans. Therefore in this book we propose the creation of a new term as an umbrella for these disciplines to facilitate communication among researchers and to create a more solid awareness among the American public of the benefits to be offered by this work. That term is, DESIGN ECOLOGY. Obviously, not all workers in all disciplines will readily accept this term. It is really being composed for the benefit of the American public, so consultants will be able to use the designation in order to be more clearly understood by a larger number of people. In a way, it is a marketing device; but in America, marketing is everything. Now we will examine some of the components covered by this new term.

FENG SHUI

By far the most ancient of the disciplines within Design Ecology is the Chinese body of thought known as Feng Shui. Several forms have been handed down, mostly through families, to the present day. Feng Shui consultants are often used in overseas Chinese communities, although the practice has been forbidden in the mainland of China under the current government. Transliterating the term Feng Shui itself into English yields the words "wind" and "water".

At its core this is a set of methods to manipulate moveable elements of design against fixed elements. Fixed elements would be such things as metal, fire and earth, while moveable elements would be analogous to air and water. All of this deals with the movement of Ch'i which can roughly be defined as a universal invisible energy creating all other powers present on this planet. Proper channeling of this fundamental energy can result in improved health, longevity, prosperity, and family harmony, according to its practitioners.

Since most of the knowledge has been passed down in families through apprenticeships, there are few fundamental texts of this art. Fortunately, the most important principles of Feng Shui are embodied in ancient Chinese writings which, when studied diligently, can give a student at least a good

8

overall background in this subject. These source books include the Tao Te Ching, by Lao Tze; The Art of War, by Sun Tzu; and the I Ching. Also, there is a profusion of almanacs published every year in Chinese communities around the world giving basic household hints involving good Feng Shui practice. Unfortunately, these are filled with cultural references which cannot be easily translated into any other language.

As with any ancient tradition, Feng Shui has gradually divided itself into several distinct schools of thought. In America the most well-known body of Feng Shui work is the Black Hat Tantric Buddhist Sect as taught by Professor Lin Yun. In his classes Professor Lin promotes a body of both material and spiritual techniques which have evolved over many centuries from diverse sources, including pre-Buddhist Tibetan Paganism (Bon Po), Buddhism, Confucianism, Taoism, and modern science. Students of Professor Lin quickly find themselves immersed in the study of a large body of techniques, many of which could be traced back to basic mental patterns. Some of the techniques could be construed as pure magic. Students who have worked diligently with his teachings have reported that they get consistent results. Since it can be difficult to find out about classes, the best current sources of information on Professor Lin's work are books by Sarah Rossbach and Elaine Jay Finster.

Other schools of Feng Shui, while not as prominent in the United States, should be considered as well in the field of Design Ecology. Most of these schools are more firmly grounded in Chinese Buddhism alone and do not tend to consider other sources as liberally as Professor Lin's Black Hat school. Evelyn Lip has written several readily available books illustrating principles from other Feng Shui schools.

From the standpoint of any Feng Shui practitioner, design is always accomplished with Ch'i power firmly in mind. Humans thrive best in an environment of balanced Ch'i – in other words not moving too fast or too stagnant. All design elements then are considered with the goal of balancing Ch'i flow, so that it moves in a moderate manner at about a speed harmonious with walking. This applies to both indoor and outdoor design.

For example, buildings should not be placed in bowl- like geologic formations because Ch'i will tend to gather and stagnate in those places and the inhabitants of buildings there will thus tend to fall ill. On the other hand, a building on a too prominent site might tend to have Ch'i flowing in and out so quickly, no one could concentrate on their work.

Another design problem illustrating the nature of Ch'i flow and its effect on people is that of moving vehicles. Any building sited in such a manner that cars are always moving towards it before turning away, creates imbalance in the inhabitants. This is because the Ch'i power of the cars will continue even past the actual physical boundaries of the car itself, and if the car appears to be moving into a space inhabitants will feel as if the car were actually moving into the space itself. The Chinese term for this kind of energy is "Sha Ch'i", which translates as a sort of death ray.

Many large volumes can be written about specific applications of Ch'i-oriented design principles in buildings alone. However, the basis of the knowledge, as traditionally practiced, should ultimately be experiential. Many Americans who have learned about Feng Shui think of it as some kind of magic. It isn't. The best results are obtained when one sees it as only a part of the hard work one must do to become successful.

While Feng Shui is a perfectly coherent discipline in itself, there are a few limitations as seen from an American design perspective. One of the first limitations is the name itself. Most people in this country do not readily identify with Chinese names. The second problem is that many modern technological artifacts are not considered in traditional Feng Shui practice. Power line electromagnetic fields, microwaves and cosmic rays were not part of the traditional Chinese world view, where invisible forces were either grouped under the concept of Ch'i or addressed in a nonspecific manner. Now, Feng Shui practitioners can and often do consider these elements in their design process, getting their information from Western sources.

A third problem with Feng Shui lies in its goal orientation. In Chinese culture the optimum goals of life were mainly centered around having a large number of prosperous descendants. When one looks at modern China and compares the current living conditions, bad as they are, with former living conditions, one could say this goal has been amply overachieved. In fact now the Chinese government must take rigorous steps to limit population growth, and this alone could account for severe restrictions on the practice of Feng Shui.

Throughout Chinese history it has usually been thought among the majority of people that immortality is conferred mainly through one's descendants. This is opposed to western concepts of immortality achieved through personal merit or individual salvation. This means in our present age, and given our cultural orientation towards somewhat different goals, we must carefully evaluate every single Chinese Feng Shui practice in light of the present ecological damage being done to this planet and the need for a universal baseline of prosperity for all humans, in order to remove the most serious threats to the survival of our species.

BAU-BIOLOGIE

This has emerged as one of the most coherent and complete disciplines covered by the term "Design Ecology".

A few Austrian and German researchers have long been aware of the influence of subtle environmental energies on humans. These energies often emanate from land forms or from underground water. Some of Bau-Biologie is associated with ancient practices of dowsing, although there are currently many attempts to build instruments to accurately measure the phenomena. Since it is in Germany that many significant advances in electromagnetic, electronic and communications technology have occurred in the past 75 years, Bau-Biologie researchers are well aware of associated field phenomena.

One of the most intriguing elements of Bau-Biologie is its consideration of the effects of ground water on humans. Apparently people who live and sleep directly above flowing ground water may develop significant health problems because of some sort of invisible "charged field" emanating from that situation. This charged field does not seem to be electromagnetic in its nature, and probably could be linked to what the Chinese call "Ch'i". Many Bau-Biologie practitioners are intimately concerned with finding the optimum place to site buildings, or in existing buildings finding the optimum place for typical living activities to occur. There have been cases reported of people moving beds from a bedroom to a living room because of an underground stream and finding that their health improves suddenly and dramatically.

An important aspect of this German body of work concerns proper building materials. This is close to part of the American practice of Clinical Ecology, which considers immune reactions to substances present in a patient's environment. Where Clinical Ecology focuses on people in medical crisis situations, Bau-Biologie is a discipline that addresses preventions of crises by using known safe materials. In this respect, Bau-Biologists are far ahead of any other researchers in the world.

If one were to concentrate exclusively on Bau-Biologie, most concepts covered in Design Ecology would eventually come forth. It is fortunate that the International Institute for Bau-Biologie & Ecology, in Clearwater, Florida, offers an excellent correspondence course in this field. This is a great way to gain important knowledge, and is highly recommended to readers of this book who want to become professionals in this field. Their certification process is better than anything else available in the United States at this time.

DOWSING & GEOBIOLOGY

In building the science of Design Ecology we should certainly consider the ancient art of dowsing as it applies to environmental situations.

We know the human body can be sensitized to subtle changes in energy patterns. Since ancient times some people have consistently demonstrated abilities to find water and sometimes resource minerals through the use of various simple techniques that all have one element in common — the use of the human body as a conduit to an indicator showing changes in energy fields or flow.

Unfortunately, right now the techniques used for measuring certain kinds of subtle energies, including ground water fields, tachyon beams, geopathic zones, and mineral emanations, are all unquantifiable. In a sense this means Design Ecologists may be stuck with old dowsing techniques until such time as instrumentation of sufficient quality and reliability is developed to measure the fields that we know are present.

During the 1991 Extraordinary Science Symposium (July 30-Aug 1) Dan Davidson presented one new form of useful instrumentation. He has developed a reliable "gravity wave detector". This detector is simply a coupling of a piezoelectric sensor through an amplifier into an ordinary voltmeter. He has tested this across several earthquake fault zones and found readings to be predictable and consistent. Designs based on this concept may soon become widely used. In addition, extremely sensitive unfiltered gaussmeters can be used to detect changes in earth magnetism. As more of us investigate correlations between magnetic patterns and other phenomena we will eventually build a coherent data base that will give us better working design principles.

Several radionic instruments may also be useful. First there comes to mind the SE-5, which might be run backwards to measure subtle energy changes in fields as the device is moved from one location to another. In general, any radionic instrument that is based on measurements taken from a probe or a well can be useful. Digital readouts are of course preferred. Unfortunately, a large number of radionic instruments are based on stick plates which are inherently a personal observation and really just an extension of dowsing in one direction. Therefore stick plate based radionic instruments must currently be rejected as a basis for building any data base of field characteristics either for an individual consultant or for general scientific use.

If, however, we have a digital readout it doesn't matter what the numbers are measuring. We can figure that out later, as long as we get numbers that change as we move through field patterns in our work, and allow comparisons between one place and another.

SYNERGETICS

Buckminster Fuller developed a new form of geometry which, as our field matures, will become increasingly useful for conceptualization and development of design parameters. Most of us are aware of Fuller's historical role as the inventor of the geodesic dome. Behind that invention was decades of serious work developing this system of geometry that not only produced this balanced stress-bearing structure, but also plots the invisible design of the Universe itself.

It is strongly recommended that we students of Design Ecology immerse ourselves in Buckminster Fuller's work. His thinking disciplines which involved consistently relying only upon one's own experiences or the experiences of reliable informants add much credibility to our work. Along the way we will find fascinating insights into the character and operation of our Universe. By correlating Fuller's insights with our own observations and practice under other disciplines we will be able to get a better idea of just what is happening in many natural subtle energy processes. This in turn will give us techniques to help repair much of the damage that has been done to our planet.

ACOUSTICS

We now know certain kinds of subtle energy which are influential in the development of natural processes such as electromagnetism and Ch'i circulation travel in a manner exactly like sound waves. We also know that there are no true vacuums in space. Thus these waves always have a medium in which to travel, despite appearances. The NASA Voyager tapes as developed by Jeffrey Thompson conclusively prove that celestial bodies emit distinctive, individually differentiated sound waves all the time. These tapes were derived from transducer data tracks on board the satellites. Each planet, moon, and ring system has a distinctive sound pattern, and some even sound like choirs of angels! Others sound like Tibetan bells, and others emit sounds which can't be compared with anything ever heard on Earth.

Music is simply a metaphor for these universal energy waves. Several good books have been written on this subject, giving ample proof that this is a universal truth. The Gandharva-Vedas of 4,000 years ago also specifically mentioned the importance of music in universal design. Ancient Chinese scriptures say much the same thing. To the Design Ecologist, this means that music, acoustical transmission, and noise levels are all legitimate parts of our practice. In fact, it is possible to originally approach a Design Ecology practice through music, and then work into other fields of study. For many, this will be exactly the path followed in the future. There is a good reason why so many people posses musical talent, outside of the Music Industry.

Some Design Ecologists could conceivably specialize in prescribing proper music for their clients. This goes beyond simple Music Therapy, into concepts that certain pieces of music could be custom-fitted to certain environments. There is a tunnel at Concourse B of O'Hare Airport in Chicago that has random tonal sequences constantly playing along its length, accompanied by soothing lights. This is a good start. Brian Eno, among others, has done a lot of experimentation in this area.

CONCLUSION

Design Ecology is a comprehensive, ever-expanding field of study. It is hoped it can be instrumental in repairing ecological damage that has been done to our planet, and in preventing future damage. It is possible to address almost any problem or crisis from this perspective, because the term lends itself to inclusiveness. A good way to define it, based on the diverse group of components we've examined here, is to say that, "Design Ecology is the process of addressing the effect of design on the physical, mental, and spiritual health of people."

PRACTICAL APPLICATIONS

We can look at Design Ecology from a number of different perspectives. Indeed, especially in the beginning, each practitioner will have a uniquely individual way of expressing this science. This is due to the fact that we are in a period of intense cross-fertilization and experimentation. Following are some of the ways in which the various disciplines constituting Design Ecology are used in practice.

FENG SHUI

This is by far the oldest discipline covered under the scope of Design Ecology. Difficulties are often encountered in its practice because of contradictory approaches promulgated by various schools. Even though its practice is currently banned in mainland China, large numbers of overseas Chinese have continued a profusion of Feng Shui schools.

We can broadly classify the contemporary schools of Feng Shui into three categories of practice. These are: Astrological, Compass (also known as Form), and Black Hat. Throughout the various schools, a system of classifications by five elements and eight situational categories (known as Ba-Gua) remains constant. This system is the core of the I Ching, and thorough study of this ancient document is highly recommended for anyone who wishes to gain competency in Feng Shui consulting. The Ba-Gua, more commonly known as Trigrams, are presented at the beginning of almost every edition of the I Ching. Most Feng Shui practice follows the "Sequence of Later Heaven" arrangement.

Where's the front door? This pretty little restaurant building has languished, mostly unoccupied, for much of the past ten years. Nobody has been able to make it work for more than a year.

The five Chinese elements are: Fire, Earth, Metal, Water, and Wood. Students of classical European thought will readily see the sharp difference between Chinese and European thinking along these lines. These elemental forces are also used extensively in medical diagnoses, and, particularly in the Black Hat schools, are directly interrelated with acupuncture practice in several key ways, including physiognomy.

Astrological schools of Feng Shui are little known in the West. This may be partly due to the confusing nature of Western Astrology, which has several schools of thought divided between different ways of reckoning the starting point of calculations. Similar problems exist among Chinese astrologers. Building correlations between Chinese and Western systems is thus extremely difficult. The symbolic structure employed in Chinese culture also remains daunting to Western students, as does the twelve-year cycle commonly used as a focal point of calculations. Chinese astrology was most intensively employed to find good matches for arranged marriages, which are often looked on with disgust in contemporary Europe and America. Therefore, little of Chinese astrology has found any practical application for Westerners.

There is a little-known system of Chinese astrology which ignores the popular 12-year cycles and concentrates on a 9-part matrix cycle of influences. This system is firmly grounded in the I Ching, but has only been used by a minority of practitioners. It's primarily known to macrobiotic people in America, under the name "Nine Ki System". It does have some value in building correlations with Feng Shui practice, because of its I Ching roots.

The Compass schools are best presented in writings by Evelyn Lip and Stephen Skinner. Emphasis is placed on physical directions and their subtle interplay with extant landforms. For example, a large, pointed mountain to the north of a building can create a "fiery" attitude among the building's occupants, causing emotional instability but imparting a sense of protection from natural disasters. If the same mountain were to the south of the building, the fiery attitude would be even more intense, with the added problem of no perceived protection from disasters.

Every Compass practitioner must become intimately familiar with the Chinese Luo-Pan compass. Several models are available, all of which contain concentric rings of Chinese characters indicating characteristics of each direction. The eight directions are classified along the lines of the Ba-Gua. Since most Chinese maps put South at the top, the "Sequence of Later Heaven" arrangement from the I Ching is the correct way to orient oneself to directional characteristics on a Luo-Pan compass.

Landforms are classified according to elemental characteristics. The associations are: Fire - pointed; Earth -flat; Metal - domed; Water - broken; Wood - columnar. These associations can also be applied to building shapes. Gradations between these shapes are given in some manuals.

Black Hat Feng Shui practitioners use the same basic conceptions as

Compass school people. Two main differences have arisen. First, instead of always relying on physical directions to determine building characteristics, the entrance of a building or room is always considered as the North end. Then, the Ba-Gua characteristics are mapped onto the room from that point. This system contains a certain amount of flexibility not found in compass-based conceptions. According to consultant Seann Xenja of Napa, California, the human mind contains an imprinted map of the Ba-Gua characteristics. This is, then, a neurolinguistic phenomenon. If a particular part of a building or room is blocked off, missing, or enhanced, behaviors will consistently occur in patterns corresponding to the observed physical characteristics.

The other difference is in the use of so-called "transcendental cures", which are a series of rituals for clearing negative energies and enhancing positive energies. These rituals are firmly grounded in Chinese culture. According to custom, they must be transmitted in a prescribed manner, and cannot be described in a book intended for the general public.

Here's another good example of a hidden front door. The family includes people with exceptional abilities, but they have had a hard time getting their business going. In this case, any good solution involves remodeling.

BAU-BIOLOGIE

Every building is considered as an individual entity composed of many material characteristics. A structured classification of building design parameters according to a specific metaphysical system is not part of Bau-Biologie, which is an important difference when compared with Feng Shui.

First, one must consider the composition of building materials, and their possible impact on health. According to Bau-Biologie, the best building materials are often those which come from indigenous natural sources. In Ireland, then, thatch roofs would be best. In southern Europe, clay tile roofs are good. Forested places provide wood. Most regions provide sufficient materials to create healthy buildings. If a region does not provide such materials, we need to seriously question the presence of humans there in the first place.

Artificially enhanced materials often are the culprits when adverse health effects are encountered. Formaldehyde and toluene are examples of chemicals used in modern building materials which have been proven to cause allergic reactions in many people, and so must be avoided. In some cases, no mitigation is possible against chemical pollution in a building.

Heating and ventilation are of great concern in Bau-Biologie. Many modern heating methods suck electrons from the air, and so weaken people in those buildings. Gas forced air and electric radiant heating are two examples of methods with this problem. The best heating systems are those which heat the floor through embedded air or liquid channels. A small amount of heat in the floor will normally provide more comfort than a large amount of air heat. Ideally, buildings need to be constructed loosely enough to allow some outside air to enter, and tightly enough to hold in sufficient heat in winter. If heating is concentrated in the floor, people will naturally feel warmer and thus less energy will be needed.

Electromagnetic fields are another concern in this discipline. No exact correlations between health effects and electromagnetic field levels are in place at this time, although anecdotal evidence provides strong suggestions of correlations at certain field strength levels. The worst health problems seem to be associated with sleeping in magnetic fields. Most Bau-Biologie practitioners simply try to bring field strengths to the lowest level possible, on the assumption that anything out of accord with natural levels is not healthy. Sometimes, this can be achieved by simply moving certain appliances around in a house. Several good books are devoted to this subject, and some are cited in the resource section at the end of this book.

Generally, everyone should be aware of the service entrance where electricity comes into a house, and keep essential activities away from that area. Magnetic field strengths can also be reduced with special metallized paints available in Europe. Demand switches, which shut off electricity to all circuits not currently needing power, are commonly recommended.

A DESIGN ECOLOGY BUILDING SURVEY

Each practitioner will develop a unique style of surveying and reporting. Some use computers to generate reports, others use paper forms, and some just rely on the client to take notes during the survey. Each approach works, as long as clients are given solid recommendations on how to redecorate or remodel for better results in one's life.

Here, I will briefly describe my own procedures, as an example of how this practice can work. First, while coming up to the door of the house, I examine the surrounding environment. Specifically, I'm looking for nearby buildings that may affect the place being surveyed, landforms such as hills, rivers, and lakes, and traffic patterns in the neighborhood. Then, the entrance is carefully considered, to make sure it's clear of any obstructions, and that it has other favorable aspects.

Once inside the house, I set up my laptop computer, usually in the kitchen or dining room, and begin taking notes. This procedure saves time, as the notes are later incorporated into a printed report for the client. Hallways, room layouts, and relative angles of walls and floors are considered. At this time, the basic furniture arrangement is checked out, especially in the living room. The entire time in the house, I always keep my nose ready for any smells. Over several years at construction sites and in offices as a temporary worker, I've become familiar with the odors of many toxic chemicals, and can identify the most commonly misused ones. Also, a basic sensing unit is used to look for concentrations of flammable gases.

After the first look around, I get out the meters. First in line is the MEDA Multi-Frequency Gaussmeter. As a backup and cross-check, I also use a custom-built unfiltered gaussmeter from Klark Kent of Super Science. All rooms of the house are surveyed with one or both meters. Then, I go back through with a Geiger counter, to look for high levels of radiation, which are rarely found. If the client is reporting health problems characteristic of electromagnetic fields, I bring out the Electro-Stress Meter, made by the Institute of Bau-Biologie and Ecology. This is a little complicated to run, because it involves about 20 feet of ground wire to get the readings right. Therefore, this test is only run in working and sleeping areas.

During the electromagnetic and radiation walk-through, I usually pick up other design factors which may have escaped initial observation. If stagnant air or ventilation seems to be a problem, I'll bring in some neutral Chinese Joss incense and do smoke tests to find air flow patterns. I also look carefully for Feng Shui considerations, such as missing corners, awkward room designs, slanted walls or ceilings, and colors. More notes are typed into the computer.

After the electromagnetic portion of the survey, I usually go outside, to look at all the landscaping. Sometimes, this also uncovers the source of any strange electromagnetic problems. At this time, I may work with dowsing for geobiological factors, if that seems appropriate based on the client's belief system and symptoms. Usually, I use a pendulum and a cheap five-band radio to uncover indications of these patterns. Sometimes compass deflections are also noted. Backup observations of structural flaws and insect activity are made.

Finally, after summarizing recommendations, usually in a priority order, I leave the site. Over the next two or three days, I may research unusual points brought out by the survey in my library and resource database. After that, the final report is edited down for printing and sent to the client, along with general information sheets and addresses of recommended suppliers. Recommendations on appropriate music to add to a place may be included.

CONCLUSION

This gives a general idea of current Design Ecology practice. Later sections in this book will go into more detail on several points presented here, including dowsing, Geobiology, and the contributions of Buckminster Fuller and Synergetics to this field.

WHY WE STUDY FENG SHUI

A strong foundation in Feng Shui principles will provide concepts useful in improving all areas of life, enhancing spiritual development, and assisting in planetary repair work. Throughout this book, the focus is on basic ideas which can be applied flexibly. Specific applications can then be observed in your own life. As you become more familiar with the foundational principles, you will be better able to create solutions to problems affecting you. Feng Shui, being the oldest stream of Design Ecology, is an excellent foundation. Then, building more modern knowledge on top of it will help you. Here, we are looking at the fundamentals of Feng Shui, and not the superficial "cures" that you can find in most books. By knowing these fundamentals, and how they relate to science, you will have a solid method of creating the best solutions to every situation, especially those details and odd problems that can't be covered in books.

MEANING OF FENG SHUI

In the Chinese language, Feng Shui literally means "Wind and Water". It has a secondary meaning of "landscape". One way to interpret the name of this science is to note that some environmental factors are fixed, and others can be easily manipulated by human intelligence. The traditional Five Elements provide a good metaphor. Fixed phenomena would include Earth, Fire, and Metal. Moveable phenomena would include Air, Water, and Wood. You could look at Feng Shui as the practice of moving flowing elements around fixed elements to optimize your environment.

PROSPERITY

This is a major concern of most people, although it takes different forms across many cultures. There are societies where wealth is defined as the ability of a man to support many wives and children. In other areas, such as the Sahel of West Africa, wealth may mean a large number of available cattle. In China, prosperity might mean lots of children to support you in old age. European cultures tend to emphasize having title to as much good land as possible. Feng Shui practice can theoretically be built around any of these goals.

General Wealth

No matter how well off most people are, it is human nature to want more. Much of Feng Shui practice in America is dedicated to the preservation and enhancement of wealth, usually as measured by money. This is what originally brings many Americans into Feng Shui practice.

Dowsing for specific items

When people are looking for specific items to enhance wealth, such as water, oil, or gold, dowsing is often used. In fact, the basic principles of dowsing as developed in Europe and America have much in common with certain Feng Shui practices.

RELATIONSHIPS

Some houses seem to attract couples who always split up while living there. Other houses always seem to hold people who stay together for a lifetime. Subtle changes in a living situation can cause large changes, by leverage, in the relationships of affected people.

A good example is a case history of an apartment in Southern California where large beams intersected both the living room and dining areas. When the leaseholder moved in, he had a wife and child. Within a few months, the wife had moved out. Subsequent girlfriends never stayed more than a few weeks, even though the man had a good personality and lots of money. A woman who was brought in to watch the apartment during the man's increasing number of road trips found that after moving in, she was unable to form any healthy relationships either. Eventually, the oppressive environment generated by the apartment caused a harmonic manifestation in the man being jailed on drug charges. In an apartment nearby with the same design, similar problems were occurring with the residents.

HEALTH

Without good health, a person has nothing. Elderly people with large amounts of money have often said they would trade all of their money in order to feel good again.

There are many ways in which Feng Shui can address health issues. By integrating ideas found by modern science with good Feng Shui practice, it is possible to help form an ideal environment for health. A good modern Feng Shui practitioner should always be on the lookout for environmental health factors such as radioactivity, powerlines, and appliances. In fact, it is quite probable that ancient Feng Shui masters had recognized radioactive and electromagnetic factors, but had called them by names not recognizable by modern students. Many Chinese practitioners had long ago warned people against living near bodies of uranium ore, while not necessarily knowing about radioactivity. This knowledge was generally gained from experience.

Now, we have Geiger counters, radios, and field meters which can give us precise measurements of these factors. All Feng Shui students should get familiar with modern environmental health measurement techniques.

More subtle patterns, as detailed by the ancients, can then be addressed. Ch'i flow may be blocked in a house, and in those cases, you will often find the residents suffering from constipation. In other cases, Ch'i could be flowing too rapidly, and residents would suffer almost constantly from invasive diseases and mental problems.

SPIRITUAL DEVELOPMENT

There are two reasons why the study of Feng Shui can assist in spiritual growth. First, you can balance your own living and working environments to create a peaceful atmosphere. If you have to constantly deal with health, relationship, and money problems, pure spiritual studies become more difficult. Clearing these problems up can help you to focus on the important spiritual issues within yourself. The second reason why Feng Shui study is spiritually desirable is that the principles themselves provide a greater insight into the workings of Nature. Feng Shui principles are graphic illustrations of the laws of the Universe in action, and can be applied in many areas.

PLANETARY HEALING

At this time in history, the most important reason for using Feng Shui is to assist in the great work of healing our home planet. Over the past few centuries, increasing misuse of the natural elements, especially Fire, has resulted in serious threats to the ability of this planet to support our species. Biblical prophecies concerning the destruction of large parts of the planet by fire are now coming to pass. An overabundance of fire has created large pollution clouds which have, in turn, increased atmospheric turbulence. Many areas of the world are now experiencing unusually violent storms as a result. If you were to look at the situation from the perspective of two thousand years ago, it would seem as if the entire planet were burning up at this point.

As participants in this material realm, we all have the opportunity to bring about healing of the planet. Feng Shui ultimately can teach us how to better cooperate with the vital invisible threads that tie ecosystems together, and give us some effective ways to accomplish vital repairs to the planetary and local environments.

PHYSICS THEORY FROM THE CHINESE PERSPECTIVE

INTRODUCTION

In China, there is a cultural framework that, when carefully combined with our own practical concepts from Western culture, will allow us to more readily comprehend this part of the Universe. This understanding could help us in many ways throughout all scientific disciplines.

Several ancient Chinese sources give us a good idea of the physics principles developed over the 6,000 year course of that civilization's history. Linguistic patterns, coupled with the extreme degeneracy of Chinese society during the late 1800's, have combined to make many of their scientific ideas appear to Western thinkers as superstitions. This does not change the scientific concepts that can be found in Chinese documents — it just makes the truth a little harder to find.

In this section, we will briefly evaluate a few original sources. This is a difficult process, because some of the sources are apocryphal in nature. Only recently have Western translators begun to compile some of this material, which is often buried in sources that could best be called folklore. Complicating a lot of this is the fact that, around 221 B.C., many of the most ancient source documents were burned, so that a lot of basic material is secondhand, from later reconstructions. The book burnings were a political phenomenon under Ch'in Shih Huang Ti, who was a great military genius but a terrible administrator. He singlehandedly swept away thousands of years of tradition in a period of less than 20 years, leaving behind remarkable works of self-aggrandizing art at Sian and a lot of confusion for later scholars.

TAO TE CHING

This document, by a legendary author, has been circulating in various forms since about 600 B.C. In the field of religion, it has been a foundational work for indigenous Chinese Taoism, Ch'an Buddhism, and the Japanese Zen tradition.

In physics, the book is valuable as a conceptual retraining tool. Many of its statements are shocking to first-time readers. They allow one to begin reforming sensory awareness patterns and means of perceiving invisible energies. The importance of "negative energy", or that which can create great accomplishment simply by not existing, is often stressed.

Throughout the book, which is written in a compact style, Tao is referred to in many ways. It is a power which forms the basis for universal harmony. It cannot be seen, tasted, touched, or smelled, and is even more powerful because of that. It can be approached only by emptying the mind of all preconceptions.

Now it is difficult to see how such lofty statements can have a practical application — until you get into quantum physics and the search for ultimate building blocks of matter. Essentially, what physicists find, the deeper they look into matter, is that there's really nothing there. That's what the Tao Te Ching has been saying all along.

On a technological level, vacuum and cavitation can be useful in developing new types of engines and pumps. About 100 years ago, John Keely built several examples of engines running on sonic principles. Resonance was often a key part of his designs. By using statements in the Tao Te Ching as a guide, it is possible to gain a greater understanding of the principles employed in these designs. Recent experiments by Dale Pond and his colleagues have shown that resonant vacuum pull is an important part of one design which has potential for effectiveness. The amount of power generated by such a design may be great enough to be extremely difficult to control.

CHUANG TZE

These are valuable expansions and commentaries on the material of the Tao Te Ching. The writings of this sage illustrate the extension of basic universal principles into the arts of statecraft and daily living. Anyone who would like to exercise their rational minds further on mysterious concepts brought forth in the Tao Te Ching should dig out this book.

From Chuang Tze, one can begin to develop a concept of matter and energy as a unified series of harmonic constructs based on an ultimate form of energy imperceptible to humans, that propagates through the Universe in a manner similar to sound. It is this form of energy, which in Chinese is termed Ch'i that produces all other forms in the Universe. See TABLE 1 for methods of rendering this word, and other essential Chinese terms, into Western languages.

CONFUCIUS

This philosopher deserves mention because of his role in correlating earlier material and putting it in a framework that Chinese people could use on a practical, everyday level. Much of his work was devoted to political and family matters, but it contains the inner physics concepts alluded to above. It is because of Confucius that many earlier works were recovered and preserved, including the I Ching.

FENG SHUI

Now we reach the heart of Chinese physical conceptions, embodied in a science they called "Feng Shui". There is no one book that contains all this science of placement and proportion. Its literal translation is "Wind and Water". Feng Shui experts have been vitally important to Chinese culture since prehistory, and they have constantly been refining their science. Most early Western sources called it "Chinese Geomancy" and then ignored it as native superstitions. Many accounts exist of colonial exasperation with the frequent refusal of Chinese workers to build railroads and buildings in the places and designs specified.

Traditionally, this knowledge is passed generationally from one expert to the next. Therefore, original Chinese writings on this subject often seem vague and scattered. It was believed that only the trained human mind could comprehend these things, and that books could confuse many key issues. Many of the current concepts are contained in a profusion of Chinese almanacs that are published every year. Most of these are never translated into any European language.

With such a dispersed, decentralized, and comprehensive system, it is no wonder that outsiders would be suspicious. Add to this the fact that often, animal symbolism was used to designate various phenomena, and you have a system that looks to the uninitiated like a mass of overly complex and stultifying superstition.

At the basis of Feng Shui is a concept that there are five primary elemental forces at work on this planet. They are: Wood, Fire, Earth, Metal, and Water. All natural and personal phenomena can be classified as some combination of these forces. Shapes, numbers, metabolic systems, landforms, and family members can all be organized under this system. From there, it is possible to observe relationships between environmental patterns and the Ch'i inherent in any person at any given time and place.

Feng Shui knowledge is largely experiential. Successful people tend to build for themselves environments containing patterns that promote beneficial Ch'i. It has been observed that this is often done on an unconscious level. Deeply troubled people gravitate to unhealthy environments containing poor Ch'i patterns. People who are studying Feng Shui begin to modify their current environments and tend to attain greater success.

By working with the experience gained from studying Feng Shui, a scientist can become familiar with how Ch'i patterns work. My own experience has shown that much of this information is inherently nonverbal, and can only be shared with others through great effort.

26

A chain of these A-frame drive-ins existed in Denver in the 1960's. They failed, as have most businesses which attempted to occupy the leftover units. A-frames concentrate the Fire element, and trying to run a business in such a building is too volatile.

NEI CHING

Individual medicine should be mentioned as it was influenced by the foundations of Chinese physics. The ancient "Book of the Insides" was the seed for much of Chinese medical practice. The conceptual framework of five primary elements is shared with Feng Shui. Ch'i concepts, when applied to medicine, create an invisible systemic framework for the human body which carries Ch'i energy. This framework can be accessed, and Ch'i flow modified, through the use of needles (acupuncture), burning herbs (moxibustion), and pressure-point massage (shia tzu). In the overlap between Chinese medicine and Feng Shui, a coherent scientific terminology may be developed. See TABLE 2 for a chart illustrating some basic correlations between the Five Elements and health characteristics.

I CHING

This foundational document is so important, the whole next section of this book is devoted to its proper use.

UNIFICATION OF THEORIES

First, we can use the concept of Ch'i energy to name the "scalar force" that so many of our researchers are struggling to define at the moment. This concept can be linked with the basic force discovered by Tesla, which Chinese practitioners traditionally said resides in magnets, and gave us the capability of generating organized vector-wave electromagnetic force (electricity). It also may solve some of the controversy surrounding Thomas Bearden's terminology. When the terms "scalar wave" or "zero point energy" are replaced with the term "Ch'i", it often becomes easier to understand what he is saying, and thus discover applications for the implied technology.

The conceptual framework of five elemental forces can assist in designing effective experiments to more beneficially channel Ch'i energy. By classifying materials (including chemical elements) according to the Chinese elemental system, we can possibly find more effective power sources and discover the harmonic frequencies of elemental combinations in catalytic reactions. These harmonic frequencies, in turn, can allow for more efficient combinations of fuel and structural materials.

Harmonics leads us to the ancient Chinese idea that music is a metaphor for the movement behind all physical reactions. This idea is shared in different forms by many cultures, but in the ancient Chinese scriptures it reached a high definition. Confucius noted that, in his time, the practice of music was degenerating, and would eventually not be a suitable tool for self-development if those trends continued. We know that ancient Chinese music was based on a five-tone scale, and each note correlated with one of the five elements. According to indications by Confucius, each piece of music would accurately describe an action of elemental forces. That's why he devoted a lot of time to warning against using music purely for personal pleasure.

For those who wish to experiment with musical principles related to the five elemental forces, it is suggested that either a flute or a wire-strung pentatonic harp (such as certain models made by Raphael in Questa, New Mexico) be used. When using a harp, be aware that damping resonating strings is as important as the note sequences.

Classifying all human experience into 64 basic situations may seem at first to be an imaginative but impractical pursuit. However, this classification may also extend to physical forces, and deserves closer study from that standpoint, because of its precise mathematical nature. Some of the transformations underlying the I Ching still have not been fully comprehended by any significant number of people, and could hold keys to better harmonization with our environment.

Western science has long been missing some basic keys to understanding how the Universe functions. Now, because of increased communication and translation with people on the opposite side of the world, we all have an opportunity to extend our understanding in ways that will create more harmonious technologies and thus a higher standard of living for all humans.

TABLE 1.
METHODS OF RENDERING ESSENTIAL CHINESE TERMS INTO ENGLISH

CH'I
WADE-GILES	Ch'i
YALE	Chi
PIN-YIN	Qi
JAPANESE	Ki
Pronunciation:	chi (soft)

TAO
WADE-GILES	Tao
YALE	Dau
PIN-YIN	Dao
Pronunciation:	dao

FENG SHUI
WADE-GILES	Feng Shui
YALE	Feng Shwei
PIN-YIN	Feng Shui
Pronunciation:	fung shwai

I CHING
WADE-GILES	I Ching
YALE	Yi Jing
PIN-YIN	Yi Jing
Pronunciation:	ee jing

NOTE: Throughout this book, the Wade-Giles system has been consistently used for all Chinese terms and names. This is the most widely used system in America, despite its flaws and the fact that it is obsolete for current international documents. The Yale system is used by some American scholars, and Pin-Yin is used for many current official purposes, having been endorsed by the mainland Chinese government.

NATURE OF THE BOOK

What is this "I Ching" that so many people consult all the time? Most people notice that it sounds Chinese, and just pass it by, with a suspicious backward glance. Others depend on it for every detail of their lives, even to the point of getting daily readings from it, as the ultimate oracle. The noted science fiction writer, Philip K. Dick, in <u>The Man in the High Castle</u>, even postulated the I Ching as a force of nature in itself. Between these two positions, a lot of confusion about the actual nature of this book is happening. Let's just say that the I Ching is absolutely vital to a solid foundation in Feng Shui.

Most people who are aware of the I Ching at all know that it is often consulted by throwing coins in a set pattern. The result of this process is called a "Hexagram", which is built up of six lines, either solid or broken. Solid lines symbolize Yang, the male principle of the Universe, and broken lines symbolize Yin, the female principle. It is also possible for lines to be of a changing characteristic, either Yang about to change to Yin or vice versa. The I Ching book itself is composed of commentaries on the 64 mathematical possibilities generated by this system of two kinds of lines in a stack of six, and the possibilities of changing lines in each case.

With its origin shrouded in legends dating back at least 6,000 years, this looks like incomprehensible, superstitious mysticism to many casual observers. The fact that many translators used terms like "good fortune", "cross the great water", and "misfortune", doesn't help much. It seems too much like fortune-telling terminology, which was not actually the idea of the original manuscript.

The first translations of the I Ching into European languages were done at a time when the work had been seriously corrupted, in the most decadent period of Chinese history ever. This was the mid-1800's, when civil authority was breaking down under misrule by an elite group of Manchurian foreigners, parts of the country were being eaten alive by European colonial powers, drug use was everywhere, and the majority of people were desperate. In this cruel and unusual environment, street fortune-tellers and psychics thrived, daily misusing this ancient work to line their pockets with the last coins of starving peasants. It is no wonder the first translators associated this work with superstition and ignorance.

Further analysis of the work convinced some scholars, most notably Richard Wilhelm of Germany, that there was something deeper in it, which was worthy of serious study and consideration. What they discovered was that the I Ching is a precise mathematical system transforming from binary code, through an octal base system, producing 64 basic situations which can categorize all possible human experiences. In a sense, you could call it the "language of the Universal computer", and anyone familiar with computer programming principles will get a tremendous amount of information by further analysis of the statements in this paragraph.

The most advanced sages of China have traditionally memorized the entire I Ching, using its concepts daily. With experience, they can tell, at any given moment, what situation from the I Ching is occurring, and at what point in a cycle the situation is. It is also possible to tell the history and future of a building by the hexagram suggested in its architectural design. Some healers are able to draw hexagrams from faces. None of these uses has anything to do with fortune-telling or oracular usage. They come closer to the original concept of the I Ching as a physics textbook.

USING THE I CHING WITH RESPECT

"Do not consult the oracle too often, or it will lose respect for you."

Most people who work with the I Ching nowadays like to consult it as an oracle, to obtain guidance for dealing with a situation. Actually, it is not necessary for one to use the I Ching in this manner to gain maximum benefits. If one is uncomfortable with oracular processes, as for example a fundamentalist Christian ought to be, that should not stand in the way of appreciation for this essential text. Much can be gained by simple reading, contemplation, and study, and many of the most advanced students of the I Ching have never consulted it as an oracle.

As mentioned before, most Americans throw three coins to obtain a hexagram when using the I Ching as an oracle. We won't detail the process here, since almost every translation gives clear instructions in the method. Several other methods of obtaining a hexagram, which should illustrate the main characteristic of the current situation, have been developed.

The oldest method of consulting the oracle, developed long before coins were minted in China, is to use 49 yarrow stalks. These plant stalks are manipulated with the fingers, being shuffled and divided several times to generate each line of the hexagram. It takes some time, which means that consulting the I Ching this way becomes a meditational process. Chinese sages have always valued the idea of comprehending everything simultaneously on intellectual and intuitive levels, and the yarrow-stalk consultation is a good way to do that.

31

TABLE 2.

ELEMENTS

(IN ORDER OF THEIR CREATION)

	WOOD	FIRE	EARTH	METAL	WATER
VIRTUE	BENEVOLENCE	PROPRIETY	SINCERITY	RIGHTEOUSNESS	WISDOM
1	DUCKWEED	OPPRESSED	OPPORTUNISTIC	CAUTIOUS	BROOK
36	BIG TREE	REGULATED	COMPASSIONATE		RIVER
72	PALM TREE	ANTAGONISTIC	SELF SACRIFICE	FIDGETY	OCEAN
VOICE	INTELLIGENCE	OUTFLOW	RHYTHM	EMBRACE	MOVEMENT
1	PARROT	SWALLOWED	JUMPY	WITHDRAWN	STAGNANT WELL
36	ANALYTICAL	PROPER	MODERATE	CONSIDERATE	DEEP POOL
72	FANATIC	FIERCE	FLAT	OVERABUNDANT	RESERVOIR
BODY SYSTEM	ANCHOR	DIGESTION	HEART	VOICE	SOCIAL
1	WEAK	CONSTIPATED	COLD	BLOCKED	SMALL
36	STEADY	SMOOTH	WARM	ALIVE	MODERATE
72	BRITTLE	ULCEROUS	EFFUSIVE	ANIMATED	TRAVELLING
SEASON	SPRING	SUMMER		AUTUMN	WINTER
SICKNESS	DEPRESSION	EXCESS JOY	OBSESSION	ANGUISH	FEAR
ORGAN	LIVER	HEART	SPLEEN	LUNGS	KIDNEYS
DESTROYS:	EARTH	METAL	WATER	WOOD	FIRE
DEST. BY:	METAL	WATER	WOOD	FIRE	EARTH

The numbers refer to relative degrees of each characteristic for that element in the human body, as diagnosed through standard Chinese protocols such as acupuncture.
1 = almost none; 36 = balanced; 72 = excessive

Other methods of consulting the I Ching are also available. Professor Lin Yun of the Black Hat Feng Shui school advocates a different method of using coins which yields a changing line every time. Ni Hua Ching, who is probably the greatest master of this book currently living, teaches a method using plant seeds, which takes a little more time than coins but less time than yarrow stalks. Some people have successfully used "bibliomancy", which means opening the book at random and using what's on that page. Of course, the more advanced students don't need to use a method of consulting the oracle, because they always know what the current situation is, and can make the best of it by using the general mathematical principles embodied in this work.

Using methods other than throwing coins to consult the I Ching is a good idea because it deepens understanding of the mathematical process underlying the book. In fact, there have been reports of people demonstrating improved mathematical ability for several weeks after using the yarrow stalk method —apparently it can unlock a subliminal mathematical process built into the human brain!

How often should one consult the I Ching? Ancient commentators universally agreed that one should use the book as an oracle as little as possible. The practice of drawing a new hexagram for each day was especially frowned upon, because one day is too little time to absorb the essence of a universal situation. It would be as if a Pagan changed the central god symbol on his or her personal altar each day, on a whim.

Ideally, one should consult the oracle only when in the most grave experience of doubt. The frequency with which one wants to consult the oracle is thus a living proof of how much doubt is residing in the soul. Put off consulting the oracle until you know you must. If doing this would conflict in any way with your spiritual practice (this is especially true for Christians and Scientologists), don't do it! No matter what your spiritual background, however, it is a good idea to study each hexagram with its commentaries, and see how they apply in world affairs and in the lives of people around you. The wisdom contained there can help you avoid making many mistakes in life.

TRANSLATIONS OF THE I CHING

Because of the ambiguity inherent in the ancient Chinese language, having more than one translation of the I Ching around is a good idea.

Originally, the I Ching was written in what's known as "Literary Chinese". This was a coded version of the Chinese language only loosely related to the spoken word. It was developed by scholars around 1200 B.C. when they saw how the spoken language kept changing all the time. This system allowed all of China to use a common written language, even when local dialects were mutually unintelligible. Actually, the system is constructed in such a way that it could be applied to any language, even English.

In Literary Chinese, everything is said with as few words as possible. Meanings are drawn from context, and if a word is repeated frequently, that has a meaning of its own beyond the pure definition of the word. Thus, when you see terms such as "great good fortune", "misfortune", "cross the great water", and other similar expressions in the I Ching, you're actually seeing translations of code words that could have widely differing meanings depending on the characters surrounding them.

This is why the issue of which translation of the I Ching is best becomes so important. The first translation into English was by James Legge, who in

1882 was absolutely bewildered by the whole project. He did the best he could, but remained convinced throughout that these people were a bunch of ignorant heathens who just needed to be converted to Christianity so they could abandon these useless superstitions. All other translations into European languages carried a similar bias until the 1930's, when Richard Wilhelm translated the book into German.

Wilhelm's translation marked a turning point in studies of Chinese religion and cosmology. For the first time, the I Ching was seen on its own merits, and separated from the corruption of Chinese society at the time. Cary Baynes translated Wilhelm's work into English in the late 1940's. This translation, known as the Wilhelm/Baynes book, is generally regarded as the most authoritative. Unfortunately, many people find the writing style thick, and difficult to relate to everyday life. If you have one, don't try reading it straight through, because you will probably not finish. Instead, regard it as a reference volume; an encyclopedia of Chinese physics.

In the 1960's, several paperback translations of the I Ching appeared. Most of them lacked a firm foundation in Chinese cosmology or culture, and in America, popularized the concept of this work as a fortune-telling tool. The I Ching Workbook by Wing is probably the best of these, but it disagrees with Wilhelm/Baynes in some key areas.

Taoist Master Ni, Hua Ching changed all this dramatically in 1983 with his translation entitled, The Book of Changes and the Unchanging Truth. As a native speaker of Chinese who has mastered English, and the heir to a long tradition of great masters, he supplied a lot of fascinating introductory material placing the I Ching firmly in its context within Chinese culture. In his masterful commentaries on each hexagram, he often includes anecdotes from legends and from his own life which serve to vividly illustrate the material. Many charts are included, which further clarify matters. That's why, for serious students of the I Ching, this is absolutely the best translation to have. Yes, it costs a little more than the paperbacks, but you will end up using it constantly as a resource, a study guide to life, and maybe, just incidentally, as an oracle.

DETERMINING BUILDING FORMS WITH THE I CHING

Each hexagram of the I Ching is a diagram of a force of Nature. A little-known method of Feng Shui depends on it as a way to determine the history and prospects of a building visually.

If you look at the face of a building, including what should be the front entrance of a building, you will see the building resolve into distinct sections.

As you do this, imagine that each section, from bottom to top, is a line of a hexagram. Does that level of the building suggest an open or closed line? With practice, it is possible to determine a hexagram for any building, and then consult the I Ching for the message being transmitted, unwittingly, by the face of the building.

This is somewhat difficult, and open to a great deal of personal interpretation. Since few people know about this technique, it is best to use it in conjunction with other Feng Shui methods, and then temper your observations with as much history of the building as you can gather.

FENG SHUI AND THE ART OF WAR

In **THE ART OF WAR** by Sun Tzu, great attention is given to the classification of terrain and situations. Interestingly enough, the categories developed in this work have good correlations to Feng Shui practice. This is because everything in the book has a metaphorical dimension. In ancient China, this was common. Writing was not to be taken lightly, as only a few elite officials and scholars knew the art. Therefore, economy of words was a virtue. The great sages strived to say as much as possible in the smallest space.

With war becoming obsolete, we now can see a valuable inner dimension to history's greatest military manual. These notes are a combination of practical Feng Shui interpretations drawn from Sun Tzu's work combined with observations drawn from typical situations in America.

CLASSIFICATION OF TERRAIN

What kind of terrain does your house or business sit in? In traditional Feng Shui manuals, animal symbolism was usually used to designate geographical forms. Many ordinary people of former eras understood this kind of symbolism, but in a more urbanized culture, this does not make any sense to most of us. Still, knowing what kind of terrain you're dealing with is important, especially when travelling or looking for a new place to live. An alternative Chinese method of classification, based on strategic needs of the military, is readily available through Sun Tzu's work.

ACCESSIBLE

The first one in the area has an advantage. Everyone else must take second place. Transportation is physically easy, but the people who arrived earlier know how to put up obstacles at will. Strategic hills can have this nature, if they are easy to climb. Often, a place just off a summit is more accessible than the summit itself. High places with a broad plateau before a summit are a good example. Many Eastern American cities are inherently this way, especially in their oldest parts. San Francisco's grandest neighborhoods also fit this description, as does Hollywood.

ENTANGLING

Easy to leave, hard to return. This describes many mountain communities. Often, a place of great physical beauty will be like this. Entanglements between people take place with extraordinary frequency. Business can be difficult due to these entanglements. For example, in Crested Butte, Colorado, a tiny community enjoyed the services of two newspapers for several years, simply because of a complex entanglement between groups of people.

TEMPORIZING

It is difficult to tell what is the best course. Both action and inaction lead to difficulty. Think of a neighborhood with tangled streets, a variety of political parties, and no real center. These areas are especially prone to deterioration. Watts in Los Angeles is an example. Even though the streets look straight on a map, once you get in there, numerous subtle geographical complications appear.

NARROW PASS

One must always approach such an area with great caution and full awareness. Any characteristic you are not aware of could cause serious problems. Here, a good example could be a stereotypical small Southern town where newcomers are automatically suspect. If you go into a place and feel like you are being constantly watched, you're in a narrow pass. In this case, you must diligently gather information in order to survive.

HEIGHT

Again, the first one there has an advantage, but transport is more difficult. In some ways, this is a combination of a "narrow pass" and "accessible" terrain. Some newer upper-class neighborhoods in America, such as parts of Berkeley and Oakland, California, fit this description. They were built on obvious heights, which looks nice, but the people who live in these places may find themselves vulnerable to all sorts of accidents, disasters, thievery, and legal disputes.

36

DISTANT

Some places are just out of the way, and there is always difficulty going back and forth. Consider parts of Nebraska. In fact, some cities have neighborhoods like this. Rancho Moreno, a suburb between Los Angeles and Riverside, California, is an extreme example. There is nothing wrong with resting in such a place, but to attempt any important action from there would be a mistake.

CLASSIFICATION OF SITUATIONS

In this section of The Art of War, classifications deal more with situations of life. This can be translated into business, family, and individual patterns. By combining the characteristics of terrain with the current situation, it is possible to get a better idea which direction to move. While all of these situations are briefly described in the original work by vague aphorisms (here paraphrased in quotes), with diligent observation the experienced Feng Shui practitioner can find them in everyday life. Then, it becomes easier to formulate appropriate strategies. Here, we are comparing these situations with common house designs.

DISPERSIVE

"Your own territory. It is not good to fight. Instead, keep a unity of purpose." Beware of complacency, which is the greatest danger when you feel secure. This is like a symmetrical box house, with its own internal balance but no good connection with the outside world.

FACILE

"Shallow penetration into hostile territory." Split-level houses exemplify this characteristic. In such a house, people often drift to cross-purposes with each other. Trailer homes, because of their narrowness, also create this feeling. Stopping at this point is a disaster. Make sure you know what you are doing, and stay connected with all parts of your organization. Then, keep working until you can change the situation.

CONTENTIOUS

"Too obviously advantageous." Everybody wants a piece of the action, and to move first is to attract shame. A house on a high hill with big windows is a relevant design to consider. Many people feel a sense of resentment towards such a house, and spurious lawsuits often result. In general, too many windows in a house creates a contentious situation. When in a place like this, keep your sense of purpose straight.

OPEN

"There is freedom of movement. Do not block others, which would cause more difficulty. Make sure all your own defenses are in place, in case anyone tries to block you." This is like a house in an ordinary street, in an ordinary neighborhood, surrounded by similar houses. Neighborhoods like this are prone to sudden, rapid deterioration. Many of the 'Rust Belt' towns of the Midwest, such as Flint, Michigan, were built this way.

INTERSECTIONS

"In a key point, find allies." This applies to any location physically good for business, where traffic passes through but can choose to stop. The key to success is to communicate, which is a whole field of business consulting work in itself. Many people who are attracted to Feng Shui are involved in business. Therefore, it is important to seek out design elements (including decorations) which invoke this sense of intersection and facilitate communication. Several ancient Chinese and Celtic motifs are useful in this regard. They can be placed inside less suitable designs as symbols of this characteristic.

SERIOUS

"The heart of a hostile country. It is best to gather in whatever you can, because lines of supply are uncertain." A house that sprawls in too many different directions, which tends to pull a family apart, creates seriousness for its inhabitants. Sprawling houses with several isolated wings have become fashionable in some newer American suburbs. People who move into new places can unwittingly find themselves in a serious situation. Use intersection motifs to dissipate problems until you can move elsewhere.

DIFFICULT

"It is hard to travel." There is really nothing one can do about the situation, except to push on through and hope things get better. A house with obvious health problems, such as sitting under a powerline or next to a chemical plant, is a good example. Occasionally, a Feng Shui professional must recommend that the best fate for a house is its destruction.

HEMMED IN

"Narrow pathways. One can be crushed easily, and so must be

constantly aware of strategic options. Do not retreat." This can be considered the direct opposite of the Intersection situation. A small house surrounded by larger ones is the obvious design example. In this case, one has to make an extra effort to catch up with everyone else in the neighborhood.

DESPERATE

"One must fight. This can occur because there has only been a small penetration into enemy territory." Beware of any community where fighting seems to be a way of life. They are on desperate ground. Awkward, sprawling house designs with sharply angled walls can breed fights among the unfortunate residents. Because of the sharp angles, a person may never feel at home. One should simply avoid these situations when possible.

REMEDIES

In Feng Shui practice, interior design elements can stimulate a chronic disposition toward one or another situation. If a situation is unhealthy, you can strive for balance by placing a new design element invoking the opposite of the situation. Sharp angles can be softened, intersection motifs can be placed if more business is desired, and sometimes pictures, mirrors, and mobiles can be employed to create an illusion of a larger, more accessible place. The possible range of remedies for these situations is as infinite as human creativity.

An odd shaped building such as this motel may attract attention, but the chaos of such a distinctive design is too often reflected in the lives of the people who must be there all the time.

FENG SHUI & BAU-BIOLOGIE:
CONFLICT, CONFUSION, OR COOPERATION?

Feng Shui and Bau-Biologie are often combined in current practice. While some people may wonder about this, blending the two disciplines actually works well. The results can be beneficial to environmental professionals and their clients.

Feng Shui has been handed down for thousands of years in China. In its homeland, most practitioners are from families which have been doing these things for centuries. There are few, if any, formally chartered schools offering any kind of diploma or certification program. This situation can make it difficult to lend credibility to the techniques. Credibility suffers even more because, when many of the terms are translated, we get things like "tigers", "dragons", "lucky shapes", and so forth. We need to understand that a predominantly rural people would use terms like this to communicate, and it is only our modern sensibilities getting in the way of understanding.

Still, Feng Shui is vitally concerned with environmental health. It ought to be recognized as the world's first environmental inspection and design discipline, so its techniques need to be reevaluated in that light if we are to truly benefit from them. Many of the techniques, especially those dealing with land forms and room arrangements, are actually mental. Sitting in an office with your back to the front door really has caused many workers to feel overly suspicious, and as soon as the desks were moved around according to Feng Shui principles, difficulties with these workers stopped.

Several books have emerged in the Western world since the 1974 publication of Feuchtwang's comprehensive anthropological study of Feng Shui. For the most part, these books are essentially compendiums of techniques, and, as such, have great value. Foremost among these has been the Sarah Rossbach book, <u>Interior Design with Feng Shui</u>. Proper application of specific techniques to various situations has often resulted in dramatic improvements. Of course, mere technique without an understanding of the underlying principles can too easily degenerate into doctrine, which can hinder actual progress.

Bau-Biologie can fill in many of the gaps left in Feng Shui practice. While its history is shorter, it can be credited with many achievements surpassing methods used by traditional Chinese practitioners. What comes to mind first is dealing with electrical pollution, which always necessitates the use of instruments to find sources.

It is nice to think that if electromagnetic fields, toxic building materials, insecticides, and other such environmental insults had been around while Feng Shui was being developed, the ancient Chinese would have addressed these issues. Instead, modern Germans happened to get there first, and created Bau-Biologie. A combination of the two disciplines can work wonders.

For example, consider a modern apartment. According to Feng Shui, the bed should be in a certain position in the bedroom, so the client can see the door. However, a magnetic field survey shows this ideal position would place your client's head in a 3.5 milliGauss field generated by the building power feed. Another location for the bed is thus demanded by a Bau-Biologie survey. Going back to Feng Shui, mirrors might be added to the room, to create a workable compromise alignment for the bed. One or the other discipline would have brought some help, but the combination created a better result.

Another recent example was a house bought as a residence for staff members of a nonprofit community group in Boulder, Colorado. Using Bau-Biologie, it was possible to find a bad mold contamination of some old wallpaper in the house. Feng Shui observations found a pattern of room arrangements which resulted in a recommendation that one room of the house be completely remodeled, because its only entrance was through a bathroom. Between the two disciplines, a comprehensive renovation plan for the house was worked out, and the owners are now pursuing it.

Nice house, but there's no clear path to the front entrance. Adding this simple element can open up a whole life.

Every discipline has its purists. There are always people who say that something must be done exactly this way, or it's no good. In the field of environmental inspecting, however, there is no good reason to arbitrarily throw away any idea because of its source. We are often dealing with people in crisis situations, and because of that, we should be ready to go with whatever works.

Therefore, combining elements of Feng Shui and Bau-Biologie can multiply the effects of our work as environmental inspectors. Going too far toward one or the other extreme could cause a client to miss out on a good solution, while being open to all possibilities creates reports which can bring tremendous benefits to our clients.

FENG SHUI OBSERVATIONS IN EUROPE

GREAT BRITAIN AND FRANCE – A COMPARISON
[Dictation 06/20/90; In the park at church St. Germain in Paris]

There is a certain skill that many of the kings and other nobility of Europe learn in developing a method of locating essential public monuments and approving the plans of roads and streets throughout their country. The responsible monarch would use these disciplines for the good of the people, although European history tells us this seems to have rarely happened. However, one good monarch seems to at times overcome the effects of many bad ones. Good examples are Frederick the Great (also a good musician), and Queen Julianna of the Netherlands.

ENGLISH CASE

London has a great deal of natural energy flowing through it. This key energy of London comes through apparently as an effect of the old Stonehenge placement. The energy of London is broad and magnanimous. It is no accident that the prime zero meridian of longitude passes through a village called Greenwich just outside of London. This is an ideal place to be an intellectual and scientific center of the Northern hemisphere, just as Jerusalem is an ideal spiritual center. London sits on some inactive fault lines, which apparently aids in its compilation of energies.

One of the most striking monuments of old London is Westminster Abbey. This is located on a rather large power spot which apparently covers the entire area of Parliament and Buckingham Palace as well. Although Buckingham Palace is on the fringe of this area, it still captures enough energy to be a viable winter residence for the current generation of the

British Royal Family. You will notice that the British Royal Family is one of the few still remaining in the entire world. It is my own opinion that they are still fulfilling their responsibility in the realm of Feng Shui practice. What they are doing is continuing to assist in the siting of public works buildings, and lending a hand wherever needed. Their ceremonies and the operation of the sacred vestments and jewels in ceremonies are significant.

This current generation of the Royal Family is presiding over a critical time in the history of the country. I don't think anyone has realized until recently just how critical this time is. The educational system is now in a state of complete collapse. Many kinds of people who have before been educated are now not obtaining any instruction of a useful nature at any time during their lives. You will observe small groups of rather loutish looking young men here and there in London. If the present deterioration of the educational system continues, we can expect these groups to grow larger and larger until they threaten the survival of the country.

Britain's Royal Family has survived due to a unique set of circumstances, including some rather enlightened monarchs at critical periods of time in its history. Britain experienced a revolutionary period a little bit before many of the remaining countries in Europe. In most of Europe, there was an Ecclesiastical reformation that threatened civil authorities only peripherally in most instances, and of course throughout Germany, the situation remained chaotic throughout the Reformation period and into most of the 19th century.

The English case, with their early civil rebellion under Oliver Cromwell, is significant. Mr. Cromwell and company, immediately upon assuming power after having executed the king, began to pawn off many of the sacred Crown Jewels. A period of great difficulty for the realm ensued. It is obvious that Mr. Cromwell, both from his background and the track of his career, knew nothing about Feng Shui, geomancy, or any other related science. He went headlong on a course leading to the rapid destruction of his power and ability to rule the country. Right after that, with the restoration of the monarch, crown jewels were immediately refashioned. It is significant that it happened so quickly, and it shows the power of this knowledge carried within that family as compared with the knowledge being carried in other families, which was not quite so powerful or strongly delineated.

Around London you can clearly see the location of Royal installations on ancient power spots. Almost every place where an important building is situated, was at one time during pre-English days a sacred spot of the old religion.

43

Two examples of this phenomenon include Windsor Castle, located high on a hill which commands a large area of view. However, it still is not quite the highest hill in the vicinity. This is an exceptional area, and a sensitive person will feel the power around Windsor Castle.

Another interesting location is the Tower of London, built on the east end of London during the time of William the Conqueror. This unique personage made a significant gesture with the building of the Tower of London. The Tower has an interesting and somewhat stable history. Its position as the ultimate repository of the sacred artifacts, such as the Crown Jewels, is significant in terms of Geobiology. The Tower of London is located on the river Thames at a distance in a harmonic interval to the location of Westminster Abbey. It balances Westminster Abbey, neatly enclosing the powerful practical energy that you will find around the city of London.

The bracketing of the energy serves to create an interesting political situation. Few people realize that the City of London itself is really only a square mile inside a greater conglomeration of boroughs which we know collectively as London. The actual incorporated City of London, which has only 6,000 inhabitants at present, is the true financial power center of England. It has been a financial power center for many generations now and it is this power which has complemented and supplemented the ability of the British monarchs to continue ruling.

The City of London, however, is technically not necessarily a part of the Royal State of Great Britain. The city of London has never officially been conquered by anybody, including William the Conqueror himself. It has always remained entirely independent, governed by its own council, with a Lord Mayor, who is certified and acknowledged by the Crown but is not appointed by the Crown.

A solid Irish monastery which has lasted over 1,000 years. It is perfectly balanced with its surroundings.

FRENCH CASE

The French case stands in sharp contrast to the British case, which offers a record of relative stability compared to other monarchies throughout the continent. This case shows a royal line that rose over a period of time, consolidated its power, ruled for a while, and fell abruptly. The fall of the French monarchy can be correlated with certain Feng Shui patterns. The building of several luxurious palaces by Louis XIV, XV, and XVI, was contradictory to principles of Feng Shui.

These kings usurped some ancient power spots for themselves and decided that it was the right of kings to do anything that came into their minds, whether it was for the good of the people or not. This has always been a recipe for political chaos. In this case, the chaos was especially deep, disturbing, and long lasting. In fact, some aspects of the chaos are still sending ripples throughout France.

Chaos began in earnest during the reign of Louis XVI. During that time, public monuments were being taken over by the king and his nobility at an alarming rate. It almost seems as if they had forgotten all they had ever known about Feng Shui and geomancy. One piece of evidence indicating

45

that these people were irresponsible monarchs is in their siting of personal palaces on places that should have been public edifices. This showed a serious contempt for the people of France. The people of France reacted in kind and the story of the Revolution is relatively well known.

The Revolution did not put in place any lasting government. The revolutionaries wandered from one building to another after having chopped off the heads of anyone who happened to complain about their activities, and in this wandering, they never did consolidate a physical seat of power. They usurped a natural power center on a plain on the right bank of Paris, which should have been an observatory area, and now contains a stolen Egyptian obelisk. In this spot the revolutionaries chose to end the lives of their real and imagined opponents. At least the obelisk is some improvement.

These revolutionaries passed quickly, and Napoleon Bonaparte replaced them. Napoleon was the epitome of the egoist man, whose ego was so large that he felt it was his duty to rule the world, not because of any ability or any particular responsibility, but because he was Napoleon Bonaparte, and that was that. The siting of his monuments reflects this. The Arc de Triomphe, a site he personally selected, is on a subsidiary high point, a perfect place for a medicine wheel. Instead he ordered a large blocky structure, which in its very construction created a tremendous amount of chaos both for himself and his regime. Now it is a traffic nightmare. A scene in the National Lampoon movie, "European Vacation", where the family is trapped in the traffic circle almost all night, seems fairly close to reality.

After the Arc de Triomphe was completed, Napoleon was already politically out of the picture, and in fact he did not get to go through his own arch until after his death. This tragedy has colored French life to this day. As a national symbol, it might also be a factor in France's growing love affair with an egoistic style of technology, which depends heavily on nuclear power.

Paris is on a strong natural power spot. It's obvious that at one time a lot of natural beauty existed there, and some of that old beauty is reflected in several architectural marvels, such as Notre Dame. In the city there is some natural beauty left, but the park areas are too small, so they compensate for the lack of park areas by planting trees along boulevards. This helps, but it is not adequate compensation for many of the residents. It is often said that a high rate of neurosis exists among Parisians, possibly worse than among New Yorkers. Extreme tension can certainly be observed in the faces of many who walk the streets of this city.

The Eiffel Tower, which is on a reflective power point, is another interesting case of monumental technology coloring its society. It is the ultimate proportional mathematical monument to technology and has no particular natural considerations. However, it was of interest to Nikola Tesla, who visited it and apparently conducted some experiments there. Since it is a construction almost entirely made of steel, it anchors Paris solidly in the age of technology.

France has now become the nation possessing the most advanced technologies on earth. Telecommunications are far superior to anybody else's in many ways. Their computer systems work very well, and everyone else in Europe seems to be trying to improve on French technological achievements. However, these achievements are not complete, and several of the state industries have to be heavily subsidized in order to survive. This is also reflected to some extent in the unbalanced Feng Shui existing in Paris, a city where a natural evolution of monarchical direction of siting public monuments was interrupted. The momentum was recovered spottily in some cases where a few people were able to come along and set up certain edifices and monuments (such as Sacre Couer on Montmarte), which do some good in terms of balancing the city. However, there are a lot of problems still to be overcome.

The French are conscious of their own problems and don't tend to welcome outside advice, so we will see an interesting case study as more new edifices are built, especially in the outlying areas of Paris where there are few power spots of interest. Powerlines chopping up all the countryside, and the greatest proportion of nuclear power plants in the world, will certainly have an effect in the future.

INVESTMENT IN A PLACE
[Dictation 06/26/90, in a small park in Paris]

In our pride we can put a lot of investment into a place that may not be viable over the long term. This can be a practical matter, such as constructing an office building in a place where nobody is going to want to stop. It can mean building a shopping center where there is not enough parking available. Sometimes it means putting together something absolutely out of accord with natural energies.

It is the things out of accord with natural energies that are most interesting to us because these failures should not have happened, and yet they do. We have seen many examples of this all over the world. You will find examples of both successes and failures around natural power spots.

What is sometimes more difficult from the perspective of a Design Ecology professional is the fact that sometimes something will be located in a place with no particular vitality one way or the other, but because of investment by the wise and wonderful people who put it together, the place succeeds in spite of itself, maybe even for a long time.

This phenomenon is interesting because investment on a spiritual level as well as on a material level can make a place into a power spot where there may not have been one before. This is fascinating when it happens, and it is something we all need to look at when it does.

In Paris, we saw ruined Roman Baths which were located in a particular spot that was great for relaxation. Somebody tried to build an Abbey there, which was a way to relax during the Middle Ages. It eventually failed. In the 1800's somebody tried to build a hotel, which failed, and now the spot is a small park on the Left Bank of Paris. It looks successful, as can see by the happiness of children playing there and the activity of young lovers who gravitate to this place to discuss their future. So this small spot has come full circle as a place to relax. Apparently, no amount of investment was able to change its basic nature.

IMPRESSIONS OF STRASBURG
[July 3, 1990 on Herrenchiemsee Island]

Strasburg is a city radiating harmony. This is accomplished to a great extent by the network of canals flowing smoothly throughout the city. These canals are designed, whether accidently or on purpose, to create a smooth flow of energy from place to place, to modulate Ch'i in attractive curve patterns.

Strasburg was well chosen to be the legislative capital of the new European Community. It is at a crossroads, and in fact the name of the city itself means Town of Roads in ancient Frankish. There are about six major highways converging here, and it is also near a highly mineralized area. The main mineral mined in the past had been silver, and there are apparently many other minerals available. There are also enough fault lines to radiate some active Ch'i throughout the area.

The overall plan of Strasburg seems to follow a natural line of the earth in many ways. There is a central plaza along a canal precisely in alignment with the Strasburg Cathedral. The central plaza is surrounded by many buildings, including the University and some of the municipal offices. The alignment of the Cathedral is very interesting because it is set on a spot that was the original Roman camp in the area. It was not a fortified hill, and not an obvious place to build a camp, but apparently recognized as a conduit of a significant amount of energy for some time. In fact, the practical considerations of the site of Strasburg's Cathedral are surprising in that the ground is soft on one side of the Cathedral, and it was not until the last century that the ground was strong enough to hold the entire structure of the south side of the Cathedral. Therefore, the tall spire of the Cathedral exists only on the north side, and one was never built on the south side. This is a classic case of a power spot existing in an area which is not necessarily a practical place to build.

COMMENTS ON HERRENCHIEMSEE: ONE OF KING LUDWIG'S PALACES

This palace is a classic example of wretched excess. King Ludwig ascended to the throne at the tender age of 18. Apparently, he had never received the full course of instruction traditionally reserved for nobility.

He built three spectacular palaces, of which Herrenchiemsee is one. Neuchwanstein, the white "fairy tale" castle crowning a hill overlooking the Rhine, is the most famous of these. Herrenchiemsee was under construction up to his death in 1886 at the age of 41. Historians estimate that he spent only about ten nights of his life in this palace. It remains unfinished, with only 20 rooms out of 70 fully decorated. At the time of his sudden and mysterious death, which may have come at the hands of political opponents, the palace was left in a state of disarray, and was eventually turned over to the state of Bavaria, which now operates it as a tourist attraction. This palace is a tour de force of the lives of the French kings Louis XIV and Louis XV, who the Bavarian king greatly admired. He often said that he longed for a return to the days when a monarch's every whim was law; this was, according to Ludwig, the natural order of things.

King Ludwig had an interesting background in that at the age of 16, midway through his studies to become a king, he heard the music of Wagner for the first time. This music left a deep impression on him. It seems to have stimulated a latent fanaticism inside his personality. This fanaticism manifested itself in the building of these palaces, in several personal eccentricities, and also in his financial support of Wagner, all in the face of opposition from almost everybody within his state. This is a classic example of the link between Wagner's music and outbreaks of fanaticism. Another example is Hitler's fascination with this particular music, to an extent that Hitler actually cultivated rare friendships with some of Wagner's descendants just to get closer to the spirit of the man.

Touring this palace is an overwhelming experience, to say the least. Fortunately, the tour is quick. If one spent very much time in this edifice, one becomes overdosed and can fall prey to quite a few physical and mental ills. The rooms of the palace are all done in 22k gold, with gilded wall panelling, statues, and a few grotesque candelabras. There is a reception chamber containing a large bed after the style of Louis XIV, who occasionally liked to hold audiences from his own bed.

The front orientation of this palace is an indeterminate west northwest direction. An edifice this large in this part of Europe should probably not be oriented in that direction. A southward orientation would be better, because people would be coming in with the beneficial energy of the sun. With an orientation to the north people are pretty much coming in with the awkward energy of some of the storms that blow across the continent. In other words, the orientation of the castle is such that disasters could more easily come to it than good.

There is a moderating influence in a large fountain set in line with the entrance. As a matter of fact it is one of the few things done right (in Feng Shui terms) in the whole construction. The large fountain is another example of wretched excess with figures of frogmen, barbarians, and princesses spitting water onto a large statue of a nude female with a child. However, the fountain does serve to break up some of the effect of the bad orientation and the long entrance pathway, and it also sends a good negative ion cloud into the region around it. However, this influence is not sufficient to offset the other negative patterns generated by the orientation and presence of this building on this island.

It also does not face any particular landmark, so one looks out from the palace down a long row of trees leading up to it, expecting to see something, and there is just little bit of open water and a nondescript patch of land on the other side. This is not a particularly good situation for anyone attempting to live there.

Paintings in the first chambers you enter depict scenes of the life of Louis XIV. In another room are paintings of King Louis XV. A grand chamber is about 200 feet long, containing 52 candelabras and 33 chandeliers. It was said that King Ludwig liked to pace up and down in this place. This chamber also is mirrored down most of its length, so you are facing mirrors most of the way through it. These opposing mirrors create a magnification of whatever personal energy one may have. There is a door on each end of this long narrow room, which would make it suitable, possibly, as a ball room. The doors at both ends tend to create a channeling of Sha Ch'i, which could be extremely destructive to a person over a period of time. If the room were filled with people, the effect would be somewhat different. They would tend to become overly energized. If King Ludwig held any receptions there they probably would have been wild affairs for the time.

If one spends very much time in that room alone, as King Ludwig apparently did during the ten days of his life he actually lived in this palace, one would probably become somewhat psychotic. Fortunately, or unfortunately, depending on your viewpoint, the construction of this palace was halted in 1886 upon the mysterious and suspicious death of King Ludwig. His magnificent constructions, including the more famous Neuchwanstein, managed to nearly bankrupt the kingdom of Bavaria. Kaiser Wilhelm was then able to take it over for inclusion in his vision of a greater Germany.

MAINZ: THE BACKWARDS CITY

Earlier, it was mentioned that in some places, directional disorientation can occur naturally, even among people with a highly developed sense. Mainz is a classic example of one of these places. On a map, the directional orientation of the city seems clear enough, but when one is actually in the streets, confusion takes hold. Many people have reported this.

Mainz is unique in that its cathedral is even built backwards. In other words, the altar is to the West, while all other Catholic cathedrals are built with the altar to the East. The only other exception is St. Peter's in Rome. It is probable that this cathedral was deliberately built this way as a conscious compensation for the odd geomagnetic field of the area. This would tend to stabilize the city.

While walking around the cathedral, several dramatic compass needle deflections were noted. This is further evidence of the unique nature of Mainz. The city is located across the Rhine from Wiesbaden, which has several hot springs. Geomagnetic field fluctuations are often associated with hot springs. Among places where these magnetic fluctuations have also been noted are near Steamboat Springs, Colorado, and Thermopolis, Wyoming.

As soon as one reaches the outskirts of town, directional stability returns. It is possible that the geomagnetic effect here emanates as a harmonic from a hot springs across the river in Wiesbaden.

GURTWEIL: A CRIME IN PROGRESS

Imagine a beautiful resort town in a picturesque valley of the Black Forest. Imagine hotels, a river, a few houses. Beautiful, right? But this particular resort has one unique feature. Somehow, it has become the crossing point for too many sets of major powerlines to be counted. It is amazing. Some houses, and even one hotel, are right under not only one set of heavy-duty lines, but in some cases even two or three sets! At least two major transformer stations have been planted in the middle of the valley, surrounded by businesses and homes. Whoever designed this travesty is surely guilty of a crime. Somebody needs to go into this town and find out what is happening. It is likely that there would be extremely high levels of cancer, immune deficiencies, headaches, and other ailments.

Of course, I'm not going in there. You couldn't pay me enough money to go into such a place for an extended period of time, to do the necessary research. Any volunteers?

ZURICH: PROSPERITY IN THE ION FIELDS

This city was built on a strong power spot. Several towers, including one shaped like a classic Hindu Lingam temple, help to anchor the energies. Some directional confusion may occur in the region around the Hauptbahnhof (main railroad station), but this quickly dissipates as one moves away from there. In fact, the physical location, including a lot of branches of the city growing down several valleys, is in itself confusing, but geomagnetically, there seem to be few problems with the senses.

Zurich is a good town for prosperity. Its location logically facilitates trade, which has taken on a worldwide scope over several centuries. There seem to be several subtle geomagnetic lines that connect with many sections of the world. Industry tends to do well.

A big problem in Zurich is where to locate freeways. Some of them have ended up in tunnels under the city. In a discussion with German Bau-Biologie expert Wilhelm Martin, we concluded that this type of construction is debilitating for the people living "upstairs". You can see apartment buildings and houses over the freeway. Flowing water is not good for humans living above it. Flowing traffic is even worse, because the electrical and geomagnetic fields generated by automobiles are not well understood at this point. We do know, however, that cars generate a tremendous amount of Sha Ch'i while moving. If you are in front of a car, that's certainly not good. Sleeping above moving cars can't be good either.

Zurich does contain several beneficial magnetic force lines which could someday be used as a foundation for alternative transportation systems. For now, the city fathers are trying their best to discourage people from using cars inside the city. These efforts often take the form of deliberately obstructing traffic at key points such as Bellevue.

One problem that Zurich shares with all mountain towns is the occasional presence of heavy ion fields as storms track through. People tend to develop mechanisms to cope with this. One of the most dramatic coping mechanisms is a very strict religion. It is no surprise that Zurich was home to Zwingli, one of the most rigid religious thinkers of all time. Now a strict religion will cause every day to be treated the same, no matter what your senses tell you. The advantage is that business can pretty much get done without unexpected interruptions. The disadvantages are many, as historical experience has proven. Citizens of Zurich seem to be looking for a better compromise with ion fields, now that they are largely aware of the situation.

MUNICH: POWER & GLORY

This city has, by European standards, only recently come into prominence. That means within the last three hundred years. Before that, not much distinguished this town from several other Bavarian cities.

There may be a moving geomagnetic fields slowly enhancing first one area, and then another in Bavaria. There are other possible explanations which we won't go into here. Munich is anchored by several ancient gates, and the location of the Marienplatz at the center of the city seems to coincide with a power spot. A healthy commercial district has grown up around there. The Marienplatz itself houses the city government, which in the time of King Ludwig was becoming increasingly more assertive. This movement was probably enhanced by the location of the Marienplatz.

Munich's greatest problem is the Foehn, which is a particularly irritating geomagnetic weather condition associated with a warm wind. It was one of the first ion cloud events to have been recognized in the world. Tourist guidebooks even warn about it now, saying that pedestrians need to be especially careful when it is going.

Compensatory mechanisms against the Foehn have not yet been adequately developed. This phenomenon may have had something to do with creating Hitler's base of support in the area. It could build up again into a nasty political problem unless the mechanisms are developed. However, work by Dr. Joseph Oberbach and other researchers may alleviate these situations in the future.

The author playing flute by the Externstine, a sacred spot in North Germany near the small city of Detmold. It has often been said that the entire text of the ancient legends called "The Eddas" is encoded in these rocks.

CIVIC FENG SHUI OBSERVATIONS IN AMERICA

Most American cities have long been in a contest to see how much alike they can all look. For several generations, New York City was the leader, and all downtown areas were to some degree or another imitating what could be found there. Now, Los Angeles is more of a leader, which is even worse, because there is no coherent look to that area. Many American cities were founded by people who were aware to some extent of geogiological energies, but since World War II the influence of these people seems to be waning in most areas. Therefore, most American cities have no newer sites which embody any consideration for natural factors. The decay of many cities could be a natural consequence of ignoring Nature's invisible formations. Following are brief comments on a few of America's cities.

NEW YORK

This is what most people in other countries pictured when they considered America up until about twenty years ago. Still, many movies are set here (although Toronto is often actually used for the filming, because it is safer and easier to shoot there).

Manhattan and Brooklyn are sacred, high-energy spots. They both have a unique combination of geography and subtle energies which make them powerful places to stand. During several of my visits to New York, I've noticed intense astral vibrations. This could be due to an accumulation of energies from the large number of people who have lived and died there over the years. Its major monuments and facilities were generally sited by people who knew what they were doing, so there was constantly an awareness of power from geobiological sources during the key formative years.

Since Manhattan is completely built out, there isn't much that can be done to degrade its basic patterns, so it remains a vital world headquarters for many activities. In the South Bronx and parts of Brooklyn, on the other hand, no attention was ever paid to correct site usage, and these areas have become chronic war zones. This situation has persisted for more than four generations, which means there are people in these areas now permanently removed from American culture. They have societies of their own, which are almost incomprehensible to most Americans. Few people are aware of this, except to know there are simply certain areas to be avoided at all costs. Intensive rehabilitation, from both the top down and the ground up, must eventually be done in these areas.

LOS ANGELES

Obvious geographical factors have made this a desirable place to live for an insanely large number of people. Also, several geobiological power spots exist in parts of the area. Chinatown seems to be located on an especially strong one, as is Hollywood. With the population explosion there, by the mid-1960's people were spreading out into less viable areas all the time. Now, places such as Rancho Moreno, Palmdale, and Sylmar are being built up, which no old-time geobiologist would ever go near. These places even look bad to the untrained eye, and building there is just asking for trouble.

Everyone seems to recognize the need for a geographic adjustment there, which will probably be accomplished by an earthquake. However, it is possible that a major disaster has been delayed because of the relatively large number of people involved in the pursuit of higher knowledge and positive goals. High consciousness can and does make the living environment better on all levels, and Los Angeles provides an ongoing laboratory. It is interesting how earth fault lines often correlate with places that have high energy for commerce and creativity. This was certainly the case in ancient Greece, Alexandria, and parts of China. If a high level of consciousness persists in any earthquake-prone region, it is theoretically possible to delay dramatic adjustments indefinitely, possibly substituting more gradual slipping.

We should, however, remember that the Native Americans in this basin had a radically different set of building practices. Due to their knowledge of the earthquake patterns, they built simple structures which could be easily reconstituted within days. Nobody would normally be hurt in such structures. Some of the groups were nomadic, and others were more settled. Before 1500, the number of people in the area may have exceeded the population of the entire rest of North America put together. What little is known of the tribes indicates that every linguistic group scattered elsewhere on the continent was represented in the area. Therefore, Los Angeles has long been a meeting ground for large numbers of people from diverse cultures, and, barring an extreme event, it will continue to be this way for many generations to come.

LAS VEGAS

Its name just means "flat place". This was a good spot to put a railroad junction a hundred years ago. However, the city is surrounded by fascinating power spots of many types. The interplay of subtle forces in the region is absolutely amazing. Its orientation follows natural lines of force. A few canyons outside town possess incredible beauty and some spots are known to be sacred to Native Americans.

WASHINGTON, D.C.

This is easily the sickest city in North America. There is a higher ratio of police officers to population than anyplace else, and yet the murder rate has continued to increase steadily. Its layout on a power spot was designed to facilitate government communications. According to the original plan, very few people were meant to live there permanently, because government was supposed to be a part-time occupation. The plan has been seriously compromised for a long time. With the advent of modern transportation, the street system has only generated confusion for most people. Political considerations have consistently overwhelmed common sense in the siting of buildings and monuments for several generations now, which compounds the problem.

The Transcendental Meditation organization pulled all its facilities out of the area in the early 1990's because of the difficult situation. In a press release, they said there was not even a small critical mass of people devoted to higher consciousness. They felt the numbers were insufficient to continue operations. Fortunately, a few other spiritual organizations have remained, so there is a chance for improvement.

I consistently recommend that people reduce dependence on the United States government as much as possible. Until sound geobiological and moral principles (and not a phony conservatism which too often means further meddling in private affairs) guide the leaders of the United States, this recommendation must stand.

WICHITA

A confluence of two rivers at a power spot near downtown make Wichita a sacred city. Native Americans long recognized it as such. Interestingly enough, the exact area of the confluence is now a Native American Center facility. There is an intense geobiological energy which has a triangular shape, correlating to the area where the two rivers meet. This shape promotes solidity. Manufacturing concerns tend to do well in the area, over the long term. Also note that Wichita has long been a center for the air industry and the military. These factors are significant, and illustrate the importance of geobiological energies in general. Two structures in the downtown area are disruptive, and are disturbing the city to some extent. Several groups have been working actively to overcome civic problems, and they are having an effect as of this writing.

DENVER

This is a grand experiment. As far as anyone know, at no time in the history of this planet has there been any civilization settled at this spot. The only thing in Denver standing from its founding in 1859 is a small part of the downtown street layout. Everything else has changed. Few old buildings exist, and civic monuments of any kind are relatively rare. However, the city has a large number of parks, which helps a lot. Some parts are built on a positive pattern, and other parts, such as the northern and southwestern suburbs, seem to cover the land virtually at random. Economically, the city has never been stable, and as of this writing the dominant employer in the area has been the United States Government and its subcontractors for the past two decades. Much of its population has traditionally been transient. So with great interest, we can all watch this experiment continue. That said, until something concrete is done to reduce the air pollution, the city will not be a particularly pleasant place. High altitude makes any air pollution much worse than it would be otherwise.

COLORADO SPRINGS

Manitou, which is said to be the original name for Pike's Peak, also can be interpreted to mean "Great Spirit". The surrounding area, and especially the Garden of the Gods, was considered highly sacred to several tribes. Legends say no war or raiding activity was to take place in the area, and peace councils were held there whenever possible. Interestingly enough, several military installations are arrayed around the city, including several key Air Force command posts. The United States Air Force is the first military organization in history to adopt the motto, "Peace is Our Profession".

PART II:

OUR INVISIBLE ENVIRONMENT

ELECTROMAGNETISM AND HEALTH

Few issues have created as much controversy in recent years as the debate about health consequences of electromagnetic fields. We know some people are profoundly affected by electromagnetic fields. Possible mechanisms for these effects have been proposed, and research shows it is plausible these fields could create disturbances at the cellular level. However, specific details of interactions are not yet known for certain by any authority. At this point, it is safe to say that sensitivity to electromagnetic fields is an individual trait, varying widely from one person to another. Therefore, a measured level which could cause great harm to one person may actually enhance health in another.

FIELD CHARACTERISTICS

Electromagnetic fields are complex. When you push electrons through a wire, a number of things happen. When you smash electrons into a screen, like a TV or computer monitor, even more things happen. Generally, three components come together in varying degrees in an electromagnetic field. First, there is a pure magnetic field, which acts in some ways like the magnetic field from a natural magnet. Second, air becomes charged with varying levels of "space charge", which is pure electricity. Third, ionization occurs, which means a number of molecules of air actually change their individual charges, sometimes hurting people by drawing electrons from the body.

MEASUREMENT

Most consultants and researchers use gaussmeters, which measure the pure magnetic field, as the main measuring tool. There are wide variations in quality among gaussmeters. One of the best currently available is the MEDA, which measures magnetic fields in three frequency ranges. Field-strength meters, which measure the amount of electric charge in the air, are important. Electrical charges in air do not necessarily correlate with magnetic fields, although either may have an impact on health. Ion detectors are rarely used, because ionization can be created by factors other than electromagnetic fields, the amount of ions created in electrical wiring is usually small, and the meters are extremely expensive.

High-frequency measurements are a whole subject in themselves. Accurate field strength levels can only be found with terribly expensive equipment usually only used by military and communications groups. Relative measurements can be gained with a special accessory antenna made in Germany and attached to a digital multimeter. Also, ordinary automobile wideband radar detectors, run with an AC power adaptor inside a building, can sometimes be useful.

SAFE LEVELS

In Europe, magnetic field safety standards of 2.5 milligauss are generally accepted by government authorities. Electric field standards vary widely from one country to another, but a level below 20 millivolts per meter (mV/m) should be safe. For most people, only personal experience will actually determine the maximum safe level. Anyone who is under a lot of external stress will want to minimize exposure to electromagnetic fields, as a precaution. The fields have a greater effect during sleep, so cutting off all electricity to the bedroom at that time can have a good effect on people.

ELECTROMAGNETIC FIELD SURVEY TECHNIQUES

Since 1990, I have been doing electromagnetic surveys of homes, offices, and properties. During that time, many fascinating situations have been encountered. This is a report from the front lines of the struggle for better health and well-being.

Plenty of controversy exists concerning the possible effects of electromagnetic fields emanating from our power system. As fans of Nikola Tesla already know, the American power distribution system is largely a series of historical accidents, and was never intended to be the final word in electrical transmission. So what we have now in this country is a standard frequency of 60 cycles per second, and several levels of step-down voltage between the time power comes off the dynamo and it gets to our homes and businesses. Each part of the process produces some wasted energy, usually in the form of electromagnetic fields that can endanger life and health.

Saying that electromagnetic fields are dangerous is considered inflammatory in some circles. However, plenty of research exists to indicate that this is at least a possibility. Even the United States Government Printing Office has made available a background paper called "Biological Effects of Power Frequency Electric and Magnetic Fields", which contains citations of many good studies. Several books available through High Energy Enterprises also give a lot of good information on the research to date. When something gets this high up in the chain of scientific evidence, it's good to pay attention. However, there's no substitute for experience in the field.

The author finding a 10.1 milliGauss reading under a powerline in Lakewood, Colorado. (Photo by Trace Hanover)

MAGNETIC FIELDS

One of my most dramatic cases involved a small newspaper office in Wyoming. A few of the employees had intermittent health problems. When these problems are impossible for doctors to track down, and they involve mysterious pains that appear and disappear with no particular pattern, then electromagnetic fields should at least be considered as a possible factor. The editor commented that he didn't like being in his office much, and then he mentioned how there had been a high turnover in his job. I asked what had happened to the previous editors. Many of them had resigned due to health problems, and at least one had died. Magnetic field levels at his desk were around 10 milligauss. The office was next door to an appliance store, and fields from several display units, including TVs and refrigerators, were coming right through the wall. He had been thinking of remodelling the building, and a less stressful spot was found for his new office. Meanwhile, he arranged to do as much work as possible in other locations.

In Kentucky, a young woman experienced severe headaches every time she came home from college for a visit. We took the gaussmeter downstairs to her bedroom, and most of the measurements were at normal background levels. The exception was at the head of the bed, around her pillow, where readings between 5 and 11 milligauss were found. Nothing in the room could account for the fields. Finally, I went around to the room directly on the other side of the wall. An attractive motorized clock, dating from around 1955, was sitting on the shelf directly behind where her bed was. We moved the clock to another spot, far away from any bedrooms, and her headaches stopped.

These two incidents point out the effects of one kind of field emanating from the power transmission system, the magnetic field. This field is an alternating current phenomenon. In DC systems, magnetic fields also occur, but usually not with the strength you find in AC systems. You can expect to find AC magnetic fields emanating from motors, transformers (like the ones used for calculators), power service feeds, and some clocks. According to the Swedish government exposure standards, which are the best-researched in the world, 2.5 milligauss (mG) is the maximum safe exposure level. Of course, some people are sensitive to lower levels, and others can tolerate much higher levels. The Swedish standard is a good guideline for clients. It can be difficult to get below that level with mitigation measures anyway.

ELECTRIC FIELDS

In an insurance office near Denver, a woman was experiencing several stress-related health problems. She reported that she never felt comfortable at her desk. An electric typewriter was on a counter next to her. It was normally used about five minutes a day, and was turned off the rest of the time. Even when turned off, it was emanating a level of 31 Volts per meter of electric field, and this was with a crude meter setup! No other significant electric or magnetic fields were found at her position. In many appliances, electricity flows to part of a transformer or motor even when turned off. This is especially true of electric typewriters and television sets. The solution is to put the offending appliance on an outlet strip, allowing you to cut the power off a few feet away from the problem area.

In America, most people who know about electromagnetic fields pay attention to only the magnetic field. (This is rarely the case in Europe.) Electric fields can often have serious health effects, since they are an actual charge in the air around us. They are difficult to measure, since the human

62

body interacts with the fields. To do accurate measurements, it is best to set up a meter with a good ground wire running outside, and use a special antenna probe. Once the meter is set up, walk away and see how the field varies. A Fluke 87 digital multimeter is excellent for these measurements, because it's possible to make a recording of maximum, minimum, and average levels.

Electric fields can be highly variable, and completely disassociated from magnetic fields. High levels can be found around some, but not all, computer monitors. Normally, the more recently manufactured monitors show lower levels. This is a problem that can be specifically addressed with a product known as the No-Rad Shield, available through many computer stores and some specialized mail-order suppliers such as Healthful Hardware of Prescott, Arizona. When installing the shield, make sure it's grounded properly in order for it to work.

Other common sources of electric fields include lamps, digital clocks, heaters, and water pipes. The last source of electric fields brings up a big problem in this country, which is grounding of electrical systems. As Spark Burmaster said, "They're trying to draft every stray piece of metal in the United States into the electrical system." Grounding is absolutely necessary as a safety measure. However, running a ground to water pipes is not always necessary, and makes people susceptible to unwanted interference from neighbors, as the pipes carry residues of all kinds of electrical operations around a neighborhood. Sometimes it is necessary to isolate the main water pipe coming into a house with a short section of non-conducting PVC.

HIGH FREQUENCY FIELDS

High frequency fields are a growing problem all over the world, as cellular phones and microwaves become more pervasive. Until recently, it was difficult and expensive to measure them. We know there can be health problems associated with high levels of these fields, at frequencies between 30 mHz and 900 gHz. This is because in the early days of microwave communications several technicians working for the phone company actually had their internal organs cooked while stepping into the paths of operating microwave transmitters. Several died within days. Recent well-publicized litigation has also indicated that problems could exist with cellular phones, which may transmit high voltage levels directly into the brain while being used.

When using cellular phones, be aware that hand-held models are potentially the most dangerous. If you have to use one a lot, transportable

63

models, which have the transmitting antenna attached to a separate battery pack unit, are safer. When phones are installed in cars with roof antennas, you're usually insulated from the actual signal by the metal roof. This means car phones are the safest.

Recently, an affordable microwave probe, which can be attached to any digital multimeter, has become available in America through the Institute for Bau-Biologie and Ecology. This German-made probe converts readings to DC volts. Typical sources of high-frequency waves include leaking microwaves, radar transmitters, cheap computers, some TV sets, and cellular phones. High frequencies travel more or less in lines of sight, especially at the higher ranges, and thus can sometimes be blocked with metal foil or metallized cloth.

How the Merkl high-frequency probe is used. (Photo by Trace Hanover)

SURVEY TECHNIQUES

When doing electromagnetic field surveys, it is important to use good equipment and prioritize your time. The common "Tri-Field" meters are easy to obtain and relatively inexpensive, but they are not very accurate and so are best suited for hobbyists. Reputable consultants in this field typically use one or two digital multimeters with special probes, a high-quality gaussmeter, and several other accessories. The amount of equipment that can be used is endless, and any aspiring consultant can be assured that whatever is purchased today will be obsolete tomorrow. Training and experience are necessary to effectively interpret measurements.

Home surveys should always start in the bedroom. This is vital, because we depend on our nightly sleep for renewal of all biological systems. To get reliable measurements of electric fields, a temporary meter ground should be run outside to a temporary copper stake whenever possible. Sometimes, a ground wire to the home system will have to do, but make sure you check the outlet first! Occasionally, I've grounded the meter to water heat radiators with passable results.

In a workplace, pay the most attention to any spot where someone needs to stay for long periods of time. Watch out for copy machines, calculator transformers underneath desks, and electric typewriters. I try to schedule office surveys during normal working hours, and still disrupt routines as little as possible. It's tricky, but this is the only way to get meaningful readings in such an environment. Home surveys are usually scheduled in the evenings, when the whole family tends to be around and electric use in the neighborhood normally is at its peak.

An array of electromagnetic test equipment.

MITIGATION MEASURES

Sometimes, simply moving appliances will take care of problems. We're lucky when we can get off that easily. In more difficult cases, turning off electrical power at night, at the circuit breaker or fusebox, will be recommended. Many people have experienced sudden, dramatic improvements in health when taking control of their electrical environment in this way. When the power feed is inconveniently located, special relays can be installed by an electrician to allow power to be switched off from a remote control by the bed.

In more difficult situations, especially those involving high frequency waves, metal foil or specially treated plastic sheeting may need to be used. As a last resort, moving out may be recommended, in the hope that the next person to live there will be one who is not so sensitive to electromagnetic fields.

Experiments are constantly being done in mitigation methods. These are highly technical, but many involve different shapes of wire. Dr. Glen Rein, while working with the Institute for HeartMath in Boulder Creek, California, a great deal of pioneering research in this area. His experiments indicated that certain wire shapes, mainly based on spirals, Moebius strips, or caduceus coils, may indeed reduce or cancel electromagnetic fields under some circumstances.

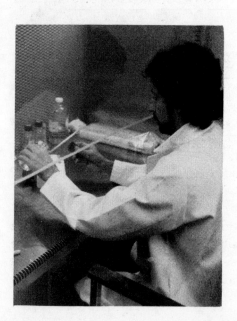

Dr. Glen Rein at work in the laboratory at the Institute of HeartMath, Boulder Creek California, October 1993.

Electromagnetic surveys are a complex issue. No matter how much you learn, there's another surprise waiting around the corner. Sometimes, fields will be found in mid-air, with no apparent source. Electrical potential in the human body is also measurable, and comparisons between people and between different locations in a building are often useful. I've learned to expect the unexpected, because our technology has gotten pretty complicated, and you never know what kind of invisible demons have sprung forth.

WHO NEEDS AN ELECTROMAGNETIC SURVEY?

Any building connected to the American 60 Hz power grid has a potential for harboring high electromagnetic field levels. According to many current researchers, any high electromagnetic field level has a potential for threatening health. There are two broad reasons why an electromagnetic survey should be ordered: health and location.

HEALTH
If someone is experiencing health problems associated with a home or office, then an electromagnetic survey should be done as part of a total environmental evaluation program. This should be done with the full knowledge and cooperation of any therapist involved.

According to several health care providers, symptoms possibly associated with electromagnetic fields include: headaches, vertigo, inability to concentrate, irritability, immune system suppression, and, in extreme cases, tumors or cancer. Since any of these symptoms can arise from a variety of causes, the therapist's role in these situations is obvious.

Three things must be considered when evaluating the role of a building in a health problem:

1. Did the problem start shortly after the association with the building began?
2. Does it go away when a person is away from there?
3. Did the symptoms fail to respond to the treatment program as expected?

LOCATION
Certain building locations ought to be checked for electromagnetic fields as a matter of course, especially in a pre-sale situation. Then, people who are known to be immune to electromagnetic fields can choose those buildings safely, and others can avoid them.

Near large powerlines

Any transmission line with large metal towers can radiate electromagnetic fields for some distance. The distance varies with line capacity, actual line usage, and weather. Measurable effects may be found up to 1/2 mile on either side of a line.

Near primary feeders

These lines often run through neighborhoods unnoticed. They are usually on slightly taller poles than regular lines, and are somewhat thicker. The wires tend to be silvery in color, with larger insulators. They are used to carry current out to smaller branches farther from generating or transformer stations. Generally, effects are seen for about 500 yards on either side.

Around transformer stations

Any unusually large electrical boxes are suspect. Stepdown transformer arrays situated in residential neighborhoods are a special problem. They have been known to create measurable fields for several blocks in any direction, depending on size. Regular residential transformer boxes are not a problem unless they're unusually close to sleeping or working areas.

Older facilities

As a rule of thumb, the older a facility, the more likely it is to be generating heavy electromagnetic fields. Even the smaller residential pole transformer cans, which are usually not a problem, can emanate distorted energy patterns when they deteriorate.

Older houses with several layers of wiring should be checked, because unusually high field levels can come from improperly connected plugs and junction boxes. Sometimes, a full electromagnetic field survey has uncovered dangerous wiring problems.

RADAR AND MICROWAVES

Houses in the path of microwave beams and radar facilities will sometimes have unusually high levels in these ranges. If a microwave or radar tower can be directly seen from the house in any direction, pay special attention to this. In some cases, such as near airports and military installations, radar pollution will pervade the entire environment for miles around. This will even show up on automobile radar detectors, which are not that sensitive.

AN ELECTROMAGNETIC ANOMALY WEST OF BOULDER

On top of Lee Hill, about 7 miles west of Boulder, Colorado, a complex of radio transmitters has evolved over a period of several years. Originally, just a sheriff's department transmitter existed, but now there are also several microwave relays and a tower for an FM station serving the Denver-Boulder area. This location has never been particularly good for the station, which experiences signal problems in many parts of the metro area. At the time it was built, several surrounding neighbors protested, to no avail. There still are bad feelings in the neighborhood about this tower, including allegations that it was illegally erected.

On January 31, 1993, I was in the area visiting at a house. I decided to take a series of readings with the Bau-Biologie Electro-Stress Meter in its space charge configuration. These readings were extremely interesting.

Outside the house, no space charge levels were detected. This is normal in a mountain location, with a low population density. As I walked into the house, sporadic readings up to 3.0 were found by walls of the house. Inside, readings were at a usual minimum of 1.0 or less. This is typical of a mountain home with wood flooring. However, near any appliances, and occasionally in midair, readings jumped up to 7.0 and more. By the TV set, the reading was about 14, which is a little high for an instant-on set when turned off. An electronic timer in the kitchen read 22, which was the highest reading in the house. In general, fluctuations in space charge readings were wider than I would expect in a rural area.

The most fascinating reading was by a metal sculpture in a hallway. No electrical circuits were present within five feet of the sculpture. However, a space charge reading of 7.0 was found extending for about a foot in any direction around it! Resonance from the radio signal could certainly be a plausible explanation for this phenomenon.

In the bedroom, another high reading up to 15 was found by a high-intensity lamp. I recommended replacing it with a regular incandescent lamp with a wooden base.

Normally, in such a house, I would expect readings to stay below 4 except by certain appliances such as the TV or a refrigerator. In almost every case where I would expect a slightly higher reading, the level was greater than expected. It is possible that these space charge fluctuations are related to the radio transmitter, which is about a mile away from the house and in direct line of sight.

SUBTLE ENERGY PROBLEMS WITH MICROWAVE OVENS

In Europe, microwave ovens are not as prevalent as in America. To some extent, this is because Europeans have simply not gotten around to buying them yet. Consumption cycles tend to be longer in Europe than in America. However, some Europeans subscribe to three lines of scientific thought which have developed there, each indicating that microwaves are not necessarily a good idea. These lines of thinking are gradually becoming more accepted on the Continent, as I discovered during a 1992 trip to Germany, Switzerland, Austria, and Netherlands. During the journey, several researchers and health practitioners were interviewed, working in a variety of disciplines. This is a summary of the current state of thinking in three scientific areas, regarding the use of microwave ovens.

RADIESTHESIA

People involved in this discipline, with a long history related to dowsing, generally agree that microwaves create disturbances in subtle energies. This opinion has been developed through large numbers of tests on food which has been microwaved. Tests usually consist of using a sensing apparatus held by an operator. Results are subjective, and are usually interpreted by the operator. The sensing apparatus can be a small tree branch, a length of copper wire, Dr. Josef Oberbach's Biotensor(TM) unit, or any number of similar devices.

Experts in this field, including Dr. Oberbach, said microwaved food has for some reason lost its nutritional potential, and, in tests, consistently shows up as being harmful to the body. Radiestheticians tend to focus on observations, without speculating as to the mechanisms involved in this nutritional loss. American practitioners of kinesiology, which is indirectly related to radiesthesia, often report similar results in tests.

MICROVITA

In this branch of science, first developed by the Indian guru P.R. Sarkar, all matter is said to be composed of particles called "microvita", which are halfway between consciousness and matter. Microvita can thus be called the building blocks of atoms. Without getting into the complexities of this theory, it can be said that, in regard to microwaved material, high-frequency radio waves cause microvita to be damaged, even though the exterior structures of atoms and molecules remain intact. When the human body takes in this food, it literally does not know how to handle it, and so may inappropriately sort out the nutrients. This can lead to problems with overweight and underweight conditions, along with other health problems, according to Microvita expert Dr. Ac. Rudreshananda. In Europe, this theory is currently being promulgated through Yoga classes conducted by the Ananda Marga organization. Experimental work supporting this theory has been designed but not yet funded.

RADIONICS

Much of this science is concerned with determining and harmonically imprinting forms of "information", defined by some of its proponents as subtle energies related to ancient concepts of ether or recent concepts of zero-point energy. Water is the ultimate carrier of information, according to radionics researchers, because it is a unique crystalline liquid. Magnetic or resonant fields can easily imprint water, according to Swiss researcher Werner Kropp. In fact, Kropp has demonstrated repeatable imprinting of water, oils, and alcohols with purely magnetic information, causing these substances to act as if they were solutions of medicines in some cases.

Since most food contains large amounts of water, it is possible to imprint information onto food just as easily as on water. Microwaves cause a random imprinting of information, which essentially causes food to become a form of "nonsense" as seen by the subtle energies of the human body. While it is possible to gain some nutrition from the food, random information may also be taken in with microwaved food, which could potentially cause harm to the body. In this line of thinking, it is the randomness of information imparted to food which is dangerous.

CONCLUSION

At this point, none of the three disciplines pointing to hidden dangers of microwave ovens is widely accepted either in Europe or America. Further research in these fields will give us a clearer idea of the extent of the danger, thus allowing people to make more informed choices in their use of cooking technologies. Most people seem unaware of the specific scientific reasoning behind possible dangers from these devices. For now, antipathy to microwave ovens in Europe remains primarily a social phenomenon.

GEOBIOLOGY

Geobiology is defined as the study of normally invisible energies emanating from the Earth, which affect people, animals, and plants. Until the 1700's, these energies were commonly recognized among Europeans. When difficulties with objectively measuring them surfaced, the concepts were abandoned by scientists. Many Chinese Feng Shui practitioners still recognize these energies, and have their own (usually secret) ways of dealing with them.

Effects of geobiological disturbances are wide-ranging, covering almost every physical and psychological problem there is. These effects seem to vary greatly from one person to another. In many cases when geobiological stress factors have been accounted for and compensated for, there have been reports of dramatic improvements.

This is one of the most controversial areas of environmental consulting. Many people do not acknowledge the existence of these energies. At the other extreme are consultants who find geobiological problems everywhere, and who are always prepared to sell you expensive remedies they've concocted. At this time, a rough classification of the possible origins of these energies has been worked out.

FLOWING WATER

These were the first geobiological energies to be found in modern times, about a hundred years ago in mountainous regions of Germany and Austria. They can be detected with expensive scintillation counters, probably because of increased neutron and neutrino emissions. Flowing water fairly close to the surface has been associated with increased cancer rates in studies dating back several decades in France, Austria, Germany, and England.

MAGNETIC DISTURBANCE

Earth's magnetic field is subject to extreme variations in some areas. Sometimes, this is associated with iron or molybdenum deposits. Large variations have been found near hot springs and volcanos. It is probable that symptoms from these disturbances are usually mental, since they could potentially affect the transmission of electricity in the human nervous system.

It has been observed that in areas of normal background magnetic fields, the field levels are masked by stronger AC fields from powerlines or transformers. In one case at Glenwood Springs, Colorado, the normal AC magnetic field associated with a working electric meter was completely suppressed, possibly by a natural magnetic field. This is a subject needing much more investigation.

MINERAL DEPOSITS

Some minerals seem to create disturbing energies when located close to the surface. Oil, radioactive minerals, crystals, and many metals have been implicated. Long-time residents of Wyoming have often warned against living over oil fields. More study and classification is needed in this area.

TACHYON PATHS

According to the late Vince Wiberg, a dowser and environmental consultant who worked in Los Angeles for over 30 years, cosmic rays constantly bombard the Earth. Their frequencies are extremely high. As they exit the Earth after some bending, their nature changes, and they are then called "Tachyons". The spots where they exit cause sleep disturbances, psychological symptoms, and occasionally illnesses. They can be found through dowsing, as no known instrumentation can pick up this kind of energy. Vince reported success in blocking these energies with bits of wire bent in G- and C- configurations. (Vince was a presenter at the 1986 Tesla Symposium, and his paper is in the Proceedings.)

Sometimes, alleged tachyon paths will be reflected in structural defects in a building. Near Cheyenne, Wyoming, there is a barn which no animals or people ever want to enter. When animals were at one time forced to occupy the barn, they literally cringed against the east wall. By the west wall, I perceived an unusually strong tachyon beam. Later, when looking back at the building, I noticed that the roofing shingles in that part of the barn looked as if they had literally been burned off, while the shingles on the east end were intact.

In another case, a fire in a house in Lexington, Kentucky started directly at a strong tachyon beam path entry. Similar phenomena have also been observed by other researchers, and often noted in private conversations.

GRID LINES

Several energy grid patterns have been worked out by various research-ers. It is known that there is some sort of invisible structure to this planet, which seems to hold it together by lines of pure energy. Buckminster Fuller's Synergetics work indicates the invisible structure in regular patterns is actually the most important of all planetary structures. Russian satellite photographs allegedly detected a pattern of large grid lines. Interestingly enough, their major intersections are often near sacred sites, such as Baboquivera in Arizona, Salt Lake City, Novgorod in Russia, and the "Bermuda Triangle".

Smaller grid systems have been proposed by Hartmann and Curry, and have been named after these researchers. Their purpose and precise effects are unknown, and they vary seasonally. Dr. Josef Oberbach has done a lot of work in documenting these energies through his Radiesthesia practice.

Living directly on a line or an intersection is said to be disturbing. In Europe, cathedrals were often located on intersections, to keep people from living on these sites, and to divert the energy to positive purposes. It is best to live slightly off the lines.

Barn near Cheyenne, Wyoming. Dark area on left end of roof is where shingles look "burned off".

DETECTING GEOBIOLOGICAL ENERGIES

These energies are generally undetectable with any standardized instruments currently available. Researchers and consultants must make do with observations gained from a variety of techniques. Dowsing, compass variations, radio static, and TV reception are all apparently affected by these energies. Dowsing with metal rods or wires is the most commonly used method of detecting geobiological patterns.

My own method of dealing with geobiological factors is to look at everything detectable by visual means and standard instrumentation first.

This involves electromagnetic and radiation surveys, and careful analysis of building design. Once all those factors have been recognized, then I look for other energies using my own perception, which has been refined through many years of experimentation and experience, along with a cheap radio and a compass.

The experimental gravity wave detector sometimes is useful, although its variations are not always enough to definitely pinpoint a disturbance.

Mostly, I like to rely on direct perception of geobiological energies, rather than the use of dowsing indicators. That's just a personal preference. Anyone can be trained to do this. It is related to the same sensory mechanism which sometimes makes it possible to see auras on people.

Always be aware that intense geobiological energies may be reflected in structural defects, such as cracks in walls or foundations, roof deterioration, and peeling paint. Once sense perceptions are found, confirming observations of these things are vital. Also look for insect infestations in areas where you have a sense of bad energies. Meanwhile, be assured efforts to develop instrumentation in this area are proceeding.

Different species of plants are also indicators of varying geobiological energies. For example, oaks are often associated with power spots in ancient folklore – good places to worship, but not to live. Any widespread fungus or parasitic mistletoe on a piece of ground is a good indication of some kind of energy pattern which should be avoided. Most fruit trees and common evergreens are good for people. As a general rule of thumb, however, one should plant any tree at least the distance of a mature branch length away from a house, because trees set any closer can damage the foundation and possibly cause water to be drawn under the house.

DOWSING TECHNIQUES AND RESEARCH

Societies all over the world are devoted to enhancing the practice of dowsing in all its forms. The biggest problem facing this field is consistent accuracy in readings. Wolfgang Maes, one of Germany's top Bau-Biologie consultants, once hired six dowsers to check an apartment which had been carefully measured by the latest (and most expensive) scientific equipment and found to be clear of any noxious energy patterns. He got back six radically different pictures of intense geobiological energy forms found by the dowsers, along with suggestions for elaborate and expensive "cures" sold by the practitioners themselves.

In spite of this kind of problem, intense experimentation persists both in the United States and Europe. A few dowsers have postulated that bits of wire can be placed in strategic patterns to block negative energies. Most of these practitioners use spirals, G- shapes, and L- shapes of pure copper wire. Since the results are still difficult to verify, all we can do is keep watching with the hope that these experiments will someday yield verified consistent results applicable in many situations.

In America, among the most effective experimenters have been the late Vince Wiberg of Los Angeles, and the team of Bill Reid and Slim Spurling in the Denver area. Vince worked as a consultant for over 30 years, and so built up a tremendous amount of experience. He has successfully taught his techniques to other individuals. Reid and Spurling work well together, and are pioneering the idea of cross-checking results as a team. They feel this creates more accuracy during the diagnostic phase of a site survey.

Slim Spurling   Bill Reid

In the future, research should focus more on this team concept, which offers promising methods of validation, in the absence of appropriate instrumentation. Comparison with gravity-wave and radionic measurements is also a good idea, and should help verify dowsing results with more accuracy. We have yet to see a fully digital radionic device that can be applied to geobiological fields, but one will probably appear soon.

For now, the basic dowsing tools for most practitioners are: hazel forks, an ancient tool rarely used today; bent wires; pendulums made from a variety of materials, including crystals, brass, and glass beads; and cheap radios, which allow dowsers to hear the effects of many subtle field patterns. The less selective a radio, the more useful it is for dowsing, because it can then receive a wider variety of static patterns, any of which could be meaningful. Recently, German researchers have reported success with using portable battery-operated UHF TV sets to find some types of forces.

ANIMALS, INSECTS, AND GEOBIOLOGY

In general, there seem to be three "polarities" of geobiological energy. These can be called direct, cross, and reverse. Animal behavior and the presence of insects have been said by European researchers to show the presence of these polarities. In fact, it was common practice in former times for farmers to allow animals to run around in areas where a house was to be built for at least a year before starting construction. They would use their observations to locate the house.

Horses, pigs, cattle, dogs, and most birds are directly polarized the same as humans, according to folklore. Therefore, any spot where these animals like to congregate would also be good for humans. The main exception to this is ponds and wallows, which are of course not good construction sites anyway. Still, these domestic animals, when not in water, do tend to congregate consistently in certain spots, and observing this can be valuable.

Cats, snakes, and goats seem to be cross polarized in relation to humans. They like places that are not too bad, but not really good for us either. If a cat normally likes to nap on a person's bed, this may be a sign of a cross polarization which could cause some long-term health problems.

Some places are absolutely polarized in a manner opposite to humans, thus stimulating the use of the term "reverse polarized". Some insects tend to favor these places. In fact, the formic acid bearing insects are an excellent indicator of places which are bad for people. This includes ants, wild bees, hornets, yellowjackets, and wasps. If hornets or wasps are constantly trying to build nests in a certain spot near one's bedroom, this is a bad sign. They tend to like certain spots, and will often build again in the same place after a nest is destroyed. Termites can also be an indication of cross- or reverse-polarized zones.

Many traditional Native American medicine people have often mentioned the role of ants in nature. Their comments indicate that ants have the job of cleaning up disordered subtle energies, in addition to their obvious role in processing natural waste materials. (Some traditions even use pebbles from anthills in religious rituals.)

A few years ago, an edition of the "Old Farmer's Almanac" printed an article on fire ants, which are the most virulent formic acid carriers among ants. They detailed the scope of the problem, which is terrible and growing in many Southern states. A key piece of information pointing to the role of these ants was that one tinkerer built a device which generated a huge magnetic field, on the assumption that, since the field was bad for animals, it might also be bad for ants, and either kill them or cause them to move out of an area. He was surprised to find that they were happy to set up housekeeping inside the device! This is further proof that the ants are in a reverse polarization to us. It is also possible that pesticides are actually making the ant problem worse, because, after the pesticide is no longer effective against the ants, they must move back to the site and try to metabolize these strange chemicals in addition to whatever attracted them there in the first place. This would make them want to stay longer!

Based on this information, the best thing we could possibly do in response to ants is to leave them alone, and let them operate. After an area is cleaned up, they will move on, on their own. If they are bothering people inside a building, then we should experiment with improving the geobiological energies first. In a few cases, I've done this in my own residence, with good results. Peter Lindemann's Spacecrafter device has been especially effective for this.

Insects said to be especially harmonious with humans are crickets and ladybugs. In fact, the Chinese customarily kept crickets as pets. Most other insects are apparently neutral in geobiological terms.

A few birds, especially screech owls, are said in Native American folklore to warn against settling in certain spots. If you have ever heard the piercing cry of a Northwestern screech owl, you wouldn't want to live around there anyway. Also, eagles, buzzards, and hawks deliberately shy away from humans. Thus, disturbing their nests in any way is very unwise, until we know more.

MODIFYING GEOBIOLOGICAL ENERGIES

Using wires shaped in G- and C- shapes, a technique developed by

Vince Wiberg, has already been mentioned. Slim Spurling and Bill Reid have gotten good results using thin copper welding rods bent in L- shapes. A few experimenters, mostly in Europe, use wires bent in spirals. If small wires can cause effects on these energies, imagine how great the effects must be from powerlines, fenceposts, electric fences, and steel buildings. Remember that according to Feng Shui, any new structure will affect other structures in the area, sometimes even several miles away. Geobiology could explain this traditional wisdom.

One area of current controversy among geobiologists is the long-term effect of deliberately blocking or redirecting these energies. Reports from the field are trickling in, and they are contradictory. Some people have reported that, after wires were placed, temporary improvements resulted. However, several weeks later, the original problems returned. In other cases, improvements were reportedly permanent. Some consultants have said in private conversations that they always pray before making any changes. There are times when a positive change for one household might adversely affect a nearby location. Discussions of spiritual topics such as Karma, Charity, and Compassion are thus common among contemporary specialists in Geobiology.

It is reasonable to expect that, in the absence of digital-readout instruments, reliable sensors, and similar accoutrements of standard environmental surveying techniques, we are going to have a lot of problems in consistently quantifying geobiological factors. At this time, we must regard all reports, whether positive or negative, as necessary data. Then, we need to find commonalities among the reports, to find out why variations in effectiveness are occurring.

Most of the modification devices marketed to people must be regarded as experimental. If anyone tells you a certain device will "definitely" improve your situation, be suspicious. Many of these devices are quite simple and inexpensive to build, often consisting of wires bent in certain shapes and enclosed in a casing. You might as well learn to build these yourself, and conduct your own experiments, given the current state of affairs. Those who are interested in this field should learn as much as possible about dowsing, to counteract the possibility of fraud. Everyone has the ability to dowse, using sensory mechanisms just beginning to be discovered by scientists.

WHAT IS RADIESTHESIA, AND DOES IT WORK?
A VISIT WITH DR. JOSEF OBERBACH

Radiesthesia is an inexact science, defined inexactly as "the study of subtle energies and their effects." The term often seems to be used as a substitute for the word "dowsing", when people want to indicate something more systematic is going on. It is enough of a buzzword in some circles that a visit with the man who is regarded as Germany's foremost Radiesthetician, Dr. Josef Oberbach, was in order.

Oberbach's qualifications in this field are legendary. He has been working for over thirty years as a practicing radiesthetician, and has several books to his credit. One of his most important books is a practice manual, which is unique in the world. This volume shows precisely how to use equipment and how to establish a discipline in interpreting results. Oberbach has also invented several devices which are said to help detect and modify subtle energies. And, even in the eyes of his detractors, he is regarded as a marketing genius. So I arranged a meeting with him at his home in a beautiful suburb of Munich.

Upon entering his house, I was immediately struck with the remarkable sense of balance in the rooms. The living room itself is a work of art, and a living representation of positive Feng Shui. A light fixture in the ceiling radiates joy and peace. The outdoor garden, visible from the living room, is a true work of art. Dr. Oberbach explained that a solarium between the garden and living room was added recently because of ozone layer destruction. He said that now, it is dangerous to be outdoors for very long because of the problem, so the solarium is a good answer.

In several areas of the living room are tachyon blockers (my term) which purportedly deflect and channel harmful rays emitted by electrical systems, underground water, and other sources. These inventions of Dr. Oberbach are somewhat similar to Vince Wiberg's technique of G-blockers, which are bits of wire used to deflect emanations from underground water and cosmic ray distortions. They are simple, and resemble one of the ancient mystical Runes. Oberbach said that this resemblance is no accident because these ancient archetypal symbols are made up of forms that have a universal nature. Psychotronics researchers would thus do well to study the forms of the runes, not for their mystical content, but for design ideas that could help accomplish specific goals, according to Oberbach.

Dr. Oberbach is "only" 84 years old. I could have sworn he was in his early 60's. His health and vitality are unquestionable. Even though he looked like he was in his 60's, his personal energy was of a man even younger than that. It was obvious to me that he's doing something right. It was not always so for Dr. Oberbach. During World War II, he was on the Russian Front, and was in Stalingrad for a time. He said, "Sometimes, I would meet a soldier, and see that he was like me, just a regular working man caught up in something we didn't understand. We would make a silent agreement; he would let me go, and I would let him go. There was no reason to kill."

Several years ago, he became very ill. Medical doctors did no good. Eventually, he realized that his problem was what Americans call "Environmental Illness". Many medical people in Germany do not want to acknowledge the existence of this phenomenon, which is a profound allergic reaction to any or all products of modern technology. Now, Dr. Oberbach maintains that everything that has been fabricated by humans carries an imprint against the energy pattern of the cosmos.

Dr. Oberbach in his home near Munich. He is holding his holding his "Bio-Tensor" dowsing instrument. On the table is one of the "Tachyon Blockers" he manufactured.

Dr. Oberbach said that conventional universities are not good places to learn anything. He had experiences in that realm which made it clear the real answers lie in finding out things for ourselves. Obviously, he is happy with his choice and his profession. Apparently, many people are now being helped by Dr. Oberbach's work. Testimonials litter his office.

Dr. Oberbach gave a demonstration of his BIOTENSOR (R) unit. He used a picture of a woman just dying, the wife of a friend, to demonstrate how the polarity of the body changes at that moment. (In this science, a representation of a thing is equivalent to the thing itself.) He showed two ends of human polarity, positive and negative. Positive polarity, in excess, results in cancerous growth. Negative polarity, in excess, results in AIDS. This deserves more explanation. According to this theory, AIDS has actually been around for a long time, but has not been specifically diagnosed because previously, people would die long before complete breakdowns occur. AIDS is thus simply the body's lack of resistance to any invasion, which correlates well with the idea that this is a result of excess negative polarity. Viruses play no role in this theory, as they are merely symptoms of unbalanced polarity. The important thing is to treat the polarity first, and then the body will figure out the best way to heal itself.

After the day of the interview, I met many people who knew of Dr. Oberbach. Some called him the "world's greatest radiesthetician". Others called him a fraud and a criminal. A journalist has taken apart one of Dr. Oberbach's household energy modulation devices and found nothing but a small piece of copper wire inside. This journalist said, "there is a kind of religion among people that you don't question these things, and you certainly don't take them apart to find out what's inside." It is true that Dr. Oberbach's devices are relatively expensive, and if all of them contain as few parts as the ones this journalist mentioned, there could be serious questions as to their effectiveness. On the other hand, it doesn't take much to manipulate subtle energies. Therefore, extensive tests will have to be run on these things.

ABSTRACT

METEOROLOGICAL: Pertaining to weather and weather forecasting.

NOMENCLATURE: Specific terms set up to describe scientific phenomena.

Three main types of invisible fields are attached to weather phenomena. The use of these classifications can assist in attaining greater accuracy in forecasting, as well as developing explanations for some of the more violent weather events.

The first, and presently most well-known class of weather fields is the electromagnetic. Under this classification would be lightning and some radio reception (DX) patterns.

The second class is ion fields. These have been documented and researched in a few scientific communities, and, while they seem tied to electromagnetic effects, they are substantially different in their origin, composition, behavior, and effects.

The third class is, for purposes of this book, called "Ch'i fields". These are the least recognized weather fields, and yet they may be the most important in their overall effects.

Conceptualization and measurement of these fields holds the promise of providing deeper insight into the origin and evolution of meteorological events.

ELECTROMAGNETICS

We are all familiar with summer storm electromagnetic phenomena. These occur world-wide, in varying degrees. Since the life cycle of typical lightning incidents is fairly well understood, no further coverage is necessary except for correlative notes.

The fact that lightning strikes are generally more prevalent over zones of high incidence of radioactive materials near the surface has been shown in satellite measurements. This is significant in terms of the theory presented here, as radioactive areas characteristically emit quantities of beta particles, which are essentially free electrons. This shows one relationship between ion fields mentioned below, and electromagnetic fields.

Less well known is the nature of radio signal bending. This correlates to local ambient electromagnetic charges in the atmosphere, and, according to this writer's own informal observations beginning in 1965, seems to bear a relationship to the amount of moisture and the turbulence present in a storm. Signals will show signs of bending consistently in any particular region over a period of time. It is probable that formal plotting of these signal patterns could lead to gauges of moisture and turbulence in storm fronts, adding to the accuracy of present estimates. These bending phenomena are here referred to as "DX patterns", after amateur radio usage.

ION FIELDS

Ion fields have often been treated as electrical phenomena, and thus are subject to the classic confusion between positive and negative that goes along with electrical terminology. This terminology states that "negative" defines a large amount of free electrons, and "positive" designates a lack of free electrons. Electrical current is generally considered to flow from negative to positive, although early observations had indicated the opposite was true. This reversal of terminology has plagued engineers, scientists, and educators ever since.

Ion fields have bioelectric effects. A so-called "negative ion field" has been shown by researchers, mostly in Europe and Israel, to be beneficial to living organisms. Apparently, living organisms use free electrons in some metabolic processes. The exact nature of these processes remains to be determined by biologists, but it is sufficient to say that, as a general rule, a relative abundance of free electrons in the atmosphere is good for living beings. A better term to replace "negative ion field" would actually be, "free electron region".

So-called "positive ion fields" have been shown to be detrimental to many humans and animals. This is because a lack of electrons in the air will cause bodies to give up electrons to the environment, thus causing stress. This stress usually shows up in apparently unrelated ways, such as an increase in clear-weather traffic accidents. Another way stress shows up is in over-reaction to equipment failures, which usually causes the equipment failure to worsen. Hyperactivity in children and irritability in adults has also been linked with these fields. Thus, a more accurate name for the "positive ion field" would actually be "electron depletion region".

It is possible that, under some conditions, electronic equipment failure could itself be precipitated by ion depletion regions. This has to do with microchip design, and could happen when an extremely small amount of current flow is disrupted by pulsating ion field regions. Not all areas of the world would have this problem. It would mainly show up in high altitude regions, such as Colorado, Switzerland, and Tibet, where ion field swings tend to be much wider than in other areas.

CH'I FIELDS

These are at once the most interesting, and most difficult fields to define. The term, "Ch'i", comes from the ancient Chinese science called Feng Shui, which is a discipline that covers passive solar design, architecture, interior design, artifact siting, and landform analysis. Although there is much controversy surrounding the proper nomenclature of this class of energy, the Chinese terminology will suffice until our understanding is better. Apparently, this energy can travel faster than light, cause immense destruction under certain circumstances, and form an impenetrable shield around any geographical area when properly controlled. It seems to propagate in a manner similar to sound waves, but can move through a vacuum. "Scalar" is another term applied to this type of field. This term comes primarily from Col. Thomas Beardon and other researchers, who have recently done much to mathematically and conceptually define this form of energy in Western scientific terms. It also may be related to Wilhelm Reich's "Orgone Energy".

Like electromagnetic and ion fields, this kind of energy also appears to "ride along" with weather changes. It is possible that Ch'i energy causes traditional medicine people (or shamans) of many tribes to describe storms in terms of distinctive personalities. In many traditions among native people of diverse areas, weather phenomena are spoken of more in terms of personality than in terms that Western scientists are familiar with. It could be that these natives are actually as right as scientists are in their descriptions — we just don't yet understand each other's languages.

New instrumentation is currently being developed that seems to measure these fields. Without going into confidential proprietary information, some of these devices that are said to measure "eloptic energy", "radionics", "etheric energy", or "molecular temperature", may in fact be measuring some aspects of Ch'i potential. If we pay attention to developments in this field, and start correlating measurement of weather phenomena using these new instruments with our usual methods, we can begin to discover more about the true nature of weather patterns.

STORM PATTERNS

For the purposes of this work, storms include a wider range of definitions than what is normally applied. Incidents of precipitation are easily recognized as storms. Also included in the general category of storm are strong wind incidents, and what we call "dry storms", which consist of strong electromagnetic patterns, wind vortexes, and little or no precipitation.

Any storm will, in general, be preceded by fields and waveforms of all three classes described in this report. Typically, a small electromagnetic charge potential, an ion depletion region, and a characteristic Ch'i field will precede any storm event. These three patterns have an infinite latitude of variation, just like the storms they precede. Then, following a storm, there will generally be little electromagnetic charge potential, an abundance of free electrons (negative ion field), and a small Ch'i pattern.

Dry storms have a significant type of pattern. They often are preceded and followed by ion field patterns similar to those seen with storms that generate precipitation. Researchers in Israel and Colorado have often observed this pattern. This can be confusing to people who are beginning the study of ion fields, as there will be times when dry storms cause inaccurate predictions.

In general, dry storms tend to be preceded by extremely intense ion depletion regions. In many cases, the intensity of observed social effects of the depletion region will be greater than with wet storms. The exception to this rule seems to come with especially intense wet storms, which have even more intense "signatures" preceding them than dry storms.

A dry storm will often be accompanied by ground and low-altitude turbulence, but may not be recognized as a storm by ground-based residents of the affected area. Aircraft pilots will certainly notice the turbulence, which may not have any explanation in an official weather report.

RADIO DX THEORY

Bending of radio signals during storm events is a study of its own. It is extremely complex, but as certain patterns emerge over a period of time, it is possible to build accurate predictive techniques around consistent locally observed phenomena.

An example of radio signal bending is the fact that, on the lower part of the FM radio band, signals typically propagate strongly over a range of 75 miles. However, on a consistent basis in 1975 through 1977, it was observed that, in Ft. Collins, about six hours before and during a storm, either in summer or winter, signals from Colorado Springs, approximately 140 miles to the south, would come in as strong as local signals. Similar phenomena can be found in any part of the world where FM signals are commonly broadcast.

Radio signal bending seems to happen most dramatically at frequencies between 20 and 450 MHz. Lower frequencies are affected in different ways, especially by solar-induced geomagnetic events, and are subject to bending over larger distances. Higher frequencies, in the microwave range, theoretically may bend relative to storm fields, and are definitely affected by solar phenomena. High frequency signal bending could reduce the reliability of radar signals under certain weather conditions, and so is of some concern to people involved in law enforcement and defense.

Coordinated observations of solar events such as flares and sunspots are being done by many researchers, along with observations of DX patterns. In fact, downloads of solar event, geomagnetic effect, and DX patterns are readily available in the United States through the National Oceanic and Atmospheric Administration's (NOAA) Web page. These can be plotted graphically with relative ease. Some critical frequency ranges are currently being mapped and even predicted relative to solar phenomena, including amateur bands at shortwave frequencies.

THEORY OF ION INTERACTIONS

Atmospheric ion theory is well accepted in some parts of the world, and totally ignored in other countries. This does not change its basic nature, which has been well-documented.

Each storm is generally preceded by an ion depletion region. This may be due to a "drawing in" of free electrons that then become part of dynamic electrical actions within the storm. The phenomenon seems to be analogous to local air currents that move toward a thunderstorm or blizzard as it approaches an area.

Storms involving precipitation are trailed by varying free electron regions. The water of the precipitation may actually carry free electrons with it into the local area, and lightning strikes could also leave numbers of free electrons behind. In any case, a feeling of well-being after the passing of a thunderstorm is an established part of folklore throughout much of the world.

The primary origin of atmospheric ions is somewhat unclear. Many particles may come from waterfalls and surf. Beta particles emitted by decaying radioisotopes may also have a lot to do with overall, worldwide atmospheric ion balance. Interestingly enough, traditional medicine people, who were charged with weather prediction, recognized certain parts of North America as being especially important in weather control. The Black Hills area, where ion-filled winds emerge from several caves, and the so-called "Four Corners" area which is home to the Hopi nation have been specifically mentioned by Leonard Crow Dog, Wallace Black Elk, Thomas Banyacya, and other spiritual leaders. Both of these areas contain significant uranium deposits.

There are probably other, more local areas of weather control scattered around each continent. One of these areas is just west of Boulder, Colorado, where there are small uranium deposits. Often, storms have been observed to change direction over the city of Boulder. That is partly why major weather research facilities are located there. The mesas above Boulder are ideal places to watch weather patterns emerge, re-form, and then move out to other areas of the continent.

It is possible that ion dynamics related to radioactivity in the soil has something to do with storm-steering. A way to test this would be to set up geiger counters over known uranium deposits, and then get strip chart readings of variations in radioactivity correlated with weather changes. Put the two together, and we can determine if these variations in radioactivity actually do have something to do with the nature and direction of storms.

PERCEPTION OF ION FIELDS

Many people have uncharted perceptive mechanisms which can be used for weather observation. It is thus possible for individuals to learn how to perceive ion and Ch'i fields for themselves. Anecdotal information about people who can "feel it in their bones" when storms are on their way is available in almost every family on Earth, and this information may in many cases have an origin in the perception of ion fields.

People who operate technical equipment can especially benefit from this kind of perception, because they can more easily prevent breakdowns by exercising extra caution at certain times. Eventually, delicate micro-circuits may have variable ion discharge components built in, which can be adjusted to ambient ion conditions to provide for safer grounding. On a more mundane level, drivers can vary their routines in response to input on ion fields, thus avoiding accidents or traffic jams.

One of the most important benefits of good ion field perception is that transient emotional problems can be kept from having more serious effects. For many people, ion fields affect emotional well-being first. When this is known, compensations can be made for the effect of ion fields, and what could have been a disaster stemming from an emotional reaction can be taken less seriously, and problems can be stopped before they get out of hand. This can even be applied to domestic disputes, which definitely increase under an ion depletion region.

Many people with arthritis or rheumatism have dramatic reactions to ion depletion regions. This may be due to a situation where free electrons being pulled from the body cause extra stress on ligament and associated muscle tissues. This, in turn, results in painful sensations. Better knowledge of the workings of these mechanisms may help to relieve the more extreme pains that some of these people suffer, and lead to a consistent electrical treatment for their conditions.

Other people have less dramatic reactions to ion depletion stress. These reactions may show up as mental phenomena, such as irritability, difficulty in concentrating, and decrease in fine motor skills. Learning to recognize these patterns as simply consistent reactions to normal weather phenomena can help people develop compensatory methods of thinking, increase work safety, and predict weather for themselves. This has far-reaching implications for people in many fields of work, especially computer operations. Any experienced data manager can tell stories about multiple breakdowns within a short period of time, and would love to be able to prevent this kind of event.

There are a few people who have no reaction whatsoever to ion fields. This seems to average out to about 10 per cent of the general population. Non-reactive people are important to society, as they can work consistently under all conditions if given the chance. Reactive people are also important, because they can see things coming that are of importance to everyone. Good communication between reactive and non-reactive people is good for timing of work flow, forecasting of weather, and generally enhanced levels of productivity.

PERCEPTION OF CH'I FIELDS

At this point, it would be valuable to look into other possible uncharted sensory mechanisms that may exist, in terms of their implications as to observation of meteorological phenomena.

Commonly, in certain circles, a phenomenon known as "auras" is discussed. These bands of light-like energy supposedly surrounding human bodies are usually said to be perceived as colors, and there are agreements among some groups as to the general meanings of these colors. Sherry Edwards has done pioneering work in the interpretation of auric patterns as sound perception. Most likely, as the existence of this phenomenon is conclusively proven, it will be as a form of Ch'i energy with its own frequency patterns relative to states of health among people.

The way a sensory apparatus that allows interpretation of auras would work is of some bearing to this discussion. According to Beardon, the human brain is a generator and receiver of the "Scalar Interferometer Effect". This effect is a phenomenon arising from the bipolar nature of the brain itself. The sensory apparaus would probably be a series of nerve pathways culminating in the pineal gland (as mentioned in Tibetan traditions), which has a location that would allow it to be a sort of mediator between the two halves of the brain. A trained observer of auras would essentially have this pathway open, and would posses an internally developed sensory reference system allowing the perceptions to be rendered into understandable language.

Naturally, there would be varying abilities among people relative to the usage of such a sensory apparatus. It stands to reason that this apparatus would be able to detect distinctive patterns in weather phenomena, and, with repeated observations and training, people could learn to interpret these patterns for those who are less skilled. This could be an explanation for the "medicine person" phenomenon observed in many cultures, where certain members of a community are credited with being able to predict weather consistently, with no sources of information from outside the immediate geographic area.

One of the most dramatic patterns, begging for interpretation, is that of a drought. Many droughts are said to be caused by shifts in a jet stream. The cause of the jet stream shift is a great mystery to scientists, although a large, stationary mass of high-pressure air is sometimes cited. Such a mass of air that would remain in one place for a long time could reasonably be assumed to have an electromagnetic component. By extension, it could be said that a Ch'i component may be involved as well. If so, someone who has the "auric sense" well developed should be able to see and interpret this standing wave, and may be able to provide information that could be useful in controlling the extent of its damage.

The same sense could also detect disastrous events as they are forming, and assist in helping people get out of the way. This conclusion is to some extent drawn from Obolonsky's work, as he proposes that, since this form of energy can bend in both time in space as it travels faster than light, we should be able to see effects of some dramatic weather events in advance.

Another sense people may have that is not well-defined in the English language is a kind of skin sensation. Commonly, there is a sensation called "the willies" that some people get when there is a non-specific source of fear. This is actually an extreme example of stimulation of this sensory mechanism. One culture on Earth seems to have this sense especially well-developed, and apparently based much of their traditional economy on it. We refer here to Australian Aborigines, and "song lines". Extensive listening to their music may cause demonstrations of this sense in sensitive people, according to private comments by Australian researchers.

Weather patterns could theoretically stimulate this skin sense, especially given what has been proposed in this book concerning the correlation of Ch'i fields with weather phenomena. Subtle skin sensations could, with further observation and statistical study, become an aid in weather forecasting.

TECHNIQUES FOR MEASURING CH'I FIELDS

Some new circuits have been shown to cause meter readings under circumstances that, to the untrained observer, should not cause any fluctuations in equipment at all. Any instrument that measures Ch'i fields is going to be difficult to evaluate. Human brains can put out intense local Ch'i fields, and have a tremendous amount of leverage in some situations relative to weather patterns. Thus, everything about the operation of such an instrument depends on the attitude of the operator. This is why many of these instruments have come under the close scrutiny of health authorities, and some instruments have been banned from commerce in the United States and other countries.

Some of the circuits that seem to be effective in these types of measurements include paired tuned-resonant arrangements of capacitors and inductors (or coils). Others actually have spiral or circular printed forms. Driving meters has been a vexing problem for some designers, as most electrical meters currently available are not very sensitive, and those that are tend to be tightly wired into circuits unsuitable for detecting these subtle energies.

Most of the available instruments which might measure Ch'i potentials require a great deal of training and experience in order to get consistent results with a particular operator. Therefore, it is imperative that anyone who is selected to use a new type of instrument in weather observations must be a person of high character and honesty. The person who trains the operator must also be of similar upstanding character. Only then are the results of observations going to have any scientific meaning. This puts a scientist in the awkward position of being a judge of moral character as well as an evaluator of data.

If so much depends on the individual operator, it would seem impossible to get the kind of consistent results from these instruments that science would demand. Some inventors are working on this problem, and we are beginning to see instruments come out which are a little less sensitive to the individual operator.

One promising method of Ch'i field measurement has recently come to light, through the work of Jack Derby, of Tucson, Arizona. He has discovered that ordinary acoustic level meters can apparently pick up these waves under certain circumstances. He is the developer of the Violet Ray Crystal Resonator, a device which can increase positive Ch'i flow in people. When the acoustic level meter probe is placed near a wire coming from the device, or near a person who is holding a probe from the device, a reading in the 90 db range is measured. The reading falls to normal as soon as the connection with the device is broken. There are a lot of implications to this, and researchers should definitely look to acoustic measurement of Ch'i forces as a real possibility.

WEATHER MODIFICATION

Here we reach a naturally controversial subject. Throughout history, technologies have evolved to attempt weather modification, usually in a small local area. As this book is being written, there is concern among many scientists and lay observers about man-made weather effects throughout the entire globe, affecting large areas through forest decimation and air pollution.

Some modification effects from human activities are obvious to most observers. Chief among these is the tremendous amount of turbulence and pollution worldwide because of the Kuwaiti oilfield fires of 1991. These have created dramatic storm events, especially in China, since May of 1991. We may need to practice some form of weather modification just to compensate for the effects of this event.

Many modern local weather modification technologies have involved cloud seeding. The people who have done this consistently have run into a strange problem. Sometimes, a local weather effect may cause a corresponding effect in a nearby location where no effect on weather was desired. In one actual case, a ski resort wanted to have more snow, and hired a weather modification company. However, a town about twenty miles downwind from the resort had no ski resort, and, while the modification program was going on, excessive amounts of snow damaged the roofs of several buildings in town. Now, if the weather modification company was successful, they are liable for the unwanted effects of their actions in the downwind town. If, however, they can't exactly prove that their efforts were successful, and the extra snowfall could be accounted for by chance, then there is no reason for the ski resort to continue hiring them. This happened in Colorado from 1979-1981, and published interviews with an official of the weather modification company are available from the author as supplementary material to this report.

Some of the ancient traditional medicine practices of tribal people were said to cause specific weather effects. In all the cases where this has been claimed, it has been said that the practitioners responsible for the effects were working with "invisible forces". Since there are recorded instances in literature and newspaper reports of these people apparently achieving results, it is probable that they are working in some way with Ch'i energy patterns.

Ch'i energy patterns are probably more amenable to human influence than the gross moisture and air interactions addressed by cloud seeding. The Scalar Interferometer Effect provides a model that makes human influence on large weather patterns theoretically possible. If a human brain can in fact transmit and receive Ch'i energies under certain conditions, and storms are largely a Ch'i phenomenon, then the possibilities become obvious. The good thing about this form of weather modification is that people can actually consider a large number of possible effects and maybe localize the weather modification in a more specific manner than cloud seeding provides.

Trevor Constable has been doing a lot of work with weather modification over the past few years. Much of his work was with resorts in Malaysia, where he attained consistent success using new physics technology to make rain during a drought. It is interesting to note that during the devastating Midwestern floods of the summer of 1993, he made a statement to the effect that this pattern was too big for anyone to change. So we may find even the most advanced weather modification techniques developed will never completely overcome natural forces.

PART III:

CHRONIC EFFECTS OF HOUSEHOLD TOXIC CHEMICALS

INTRODUCTION AND DEFINITIONS

We could say that a toxic chemical spill may have happened in every home in the United States, and this is the result.

Some people are much more sensitive to chemicals than others. Those who are sensitive often demonstrate a variety of symptoms known collectively as "Environmental Illness" (EI). This is admittedly a controversial medical phenomenon, but for the purposes of this section its validity will be assumed. EI can advance to a state known as "Multiple Chemical Sensitivity" (MCS). In this condition, people become allergic to a wide range of chemicals associated with modern life, and must isolate themselves from our society in order to recover.

Several dangerous chemicals and chemical families commonly found in homes are summarized with regards to usage and known chronic effects as related to EI and MCS. This is organized as a guide to substances by the most well-known chemical family name. While this may seem a bit technical, more common usages and simpler names are found within each section. Words which are not defined as you go in this section may be found in the Glossary at the end of this book. If not found there, they should be in any good dictionary.

It is difficult to discuss hazardous chemicals without getting technical, partly because of all the trade and brand names used to keep people from thinking about what we're really using. The technical names and formulas are, unfortunately, the only consistent way we have to track some of these substances. To make things even more complicated, it has often been estimated that over 8 million man-made chemicals are currently being used on this planet. Nobody can keep track of them all any more!

Asbestos is specifically not covered in this work, since so many resources are already available on that subject. In homes, it is usually only dangerous when crumbling.

Many of the chemicals have listed values for TLV and IDLH. These are technical terms commonly used among hazardous waste handlers, to give a rough idea of how toxic a certain chemical might be. Here are the definitions for some of the most baffling technical terms used here:

TWA: "Time-Weighted Average", an arbitrary measurement determined by US Government laboratories to be the maximum amount someone should be exposed to over a period of time.

TLV: "Threshhold Limit Value", which usually means the maximum amount someone should be exposed to over a period of 8 hours.

IDLH: "Immediately Dangerous to Life and Health", which should be self-explanatory.

ppm: Most measurements are given in "parts per million" (in air).

All these values were arbitrarily determined, and are used as guidelines by manufacturers, waste handlers, and transport personnel. However, they may be too high for many people, especially in the average home environment.

GENERAL CHARACTERISTICS OF CHEMICAL HAZARDS IN THE HOME

Two main pathways exist for invasion of homes by toxic chemicals. Cases of pollution from nearby industrial processes exist, and have been documented in news reports, as in the ASARCO lawsuit in North Denver or the infamous Love Canal case. These incidents are complex legal problems, which usually take years to resolve in courts. Expert testimony is always needed, and usually the paperwork generated amounts to huge truckloads. In this book, we will generally not be concerned with these cases, except to encourage you to be aware of what's going on in your community, and keep in touch with your neighbors. These two steps could greatly reduce the amount of misery generated by intentional or unintentional chemical assaults.

A more common path of home contamination originates in the supermarket. Daily, hundreds of chemicals are purchased in order to accomplish the most ordinary household functions. Most people never give this a second thought. However, some people must constantly be on guard for even the slightest presence of certain chemicals. They are called, "Environmentally Ill".

Some of the more dangerous chemicals used in the home are typically cleaners. Something so mundane is understandably a source of great controversy. After all, everyone must clean house at some point. Fortunately, only a few chemicals used in cleaning are currently considered toxic by most authorities. Several manufacturers have already started to eliminate the most suspect substances in favor of possibly less effective but probably more benign alternatives. A few companies, such as Earth Wise and Seventh Generation, devote their entire production to nontoxic cleaners. It is commonly understood that the traditional, less toxic ingredients used in these formulations are not always as effective for their intended use as modern chemicals, but for some consumers, the trade-off is certainly worthwhile. A little extra work in return for better health is a pretty good deal.

CLINICAL EVOLUTION OF ENVIRONMENTAL ILLNESS

SENSITIZATION PHASE
Definition

Nobody has yet come up with a universally accepted definition for "Sensitization". Each authority seems to say something different, and some medical doctors, during informal personal interviews, have steadfastly refused to acknowledge the phenomenon even exists. Still, something is happening to some people, and it seems to be related to chemicals. If it is purely psychological, as some medical experts have maintained, the evidence for that position is lacking among the sufferers, who usually get better only when removed from the presence of chemicals.

For the purposes of this book, we will assume that the phenomenon of Sensitization exists. Its definition in this context, then, is as follows:

SENSITIZATION A process whereby a person gradually or suddenly demonstrates reactions to environmental factors which had previously been tolerated. This is usually seen in the context of man-made chemicals.

Someone who demonstrates sensitivity to chemicals is said to have "Environmental Illness" (EI). Its definition varies from one therapist to another, but we can say that it implies degrees of intolerance for many man-made substances. Within the past five years, a condition called "Multiple Chemical Sensitivity" (MCS) has been named and recognized by Clinical Ecologists and Allergists. It can be defined broadly as a cluster of chemical sensitivity reactions to more than one chemical. It probably involves the entire immune system.

<u>Megadoses</u>

Many Environmentally Ill patients were at one time exposed to a large dose of a particular chemical, usually in an occupational context. Interviews with patients and surveys of reference material have shown the following single chemicals to have triggered subsequent intolerances, according to the patients themselves: EDB, Trichloroethylene, DDT, Phenol, and Malathion. In addition, some people have reported EI or MCS to be triggered by Formaldehyde or 4,Phenyl Cyclohexene, although it has been difficult to see how these chemicals, which are considered relatively mild, could have been single triggering factors. Many other industrial process chemicals and insecticides have also been mentioned as possible sensitizers over the years.

Other cases were apparently triggered by long-term exposures to single chemicals or mixtures, which accumulated to a large exposure over time. Among factors mentioned in this context by interviewees are cigarette smoke (which contains Formaldehyde, Benzene, and Acrolein, among many other chemicals), Phenol, and Carbon Tetrachloride.

<u>Co-factors</u>

As in many modern illnesses (including AIDS), co-factors have a lot to do with the development of EI and MCS. Many cases certainly exist where someone acquired EI, while a co-worker in the same environment, exposed to the same chemicals, never developed a problem. Poor diet and lack of exercise are common stress factors for many Americans. Too many workers are exposed to intense emotional stress on the job, by cruel or careless supervisors. Genetic makeup is probably a significant co-factor as well. These co-factors account for the frustrating tendency of EI and MCS to be impossible to analyze and quantify to the satisfaction of many medical practitioners.

Just as it is rare for an EI patient to be sensitive to only one chemical, so is it rare for the origin of the illness to be associated exclusively with one chemical. Medical personnel who accept the existence of EI tend to speak of it in broad terms, instead of as a series of isolated factors.

Several clinicians, all working independently, have stated that EI and MCS may result from over-stressing the immune system, to a point where it can no longer "supervise" proper metabolization of chemical intruders to the body.

SENSITIVITY PHASE

Symptoms

Typically, most EI patients demonstrate a wide range of symptoms which tend to be non-specific. Among the most widely reported are: upper respiratory discomfort, headaches, severe mental disturbances, skin rashes, chronic fatigue, loss of appetite, itching and burning of eyes, seizures without normal electrical signatures, and asthma attacks. If no conventional treatment results in abatement of the symptoms, and no organic cause can be found, then it could be said that Environmental Illness is the relevant diagnosis.

Multiple Chemical Sensitivity is usually found subsequent to a diagnosis of Environmental Illness. Its characteristics include violent, allergy-like reactions to the presence of any one or combination of chemicals, even in amounts too small to be detected with conventional equipment such as Photo-ionization Detectors.

Repeat exposures

After sensitization, repeat exposures to the original sensitizing chemical tend to create violent reactions. Removal of the offending chemical will usually cause the symptoms to abate, which is sometimes the only way anyone can know that this is a sensitivity problem. Skin patch tests sometimes can isolate a chemical sensitivity, but are virtually useless and even dangerous in cases of Multiple Chemical Sensitivity. Usually, the repeat dose causing a reaction is infinitesimal compared to the original dose that caused sensitization.

New sensitivities

If exposure to the offending chemical continues, which is often the case, especially in workplace situations, new chemical sensitivities begin to develop. This is a part of EI often regarded as "weird" when medical personnel are called in.

As an example, one interviewee for this section said that after her exposure to EDB, a now-banned pesticide, while on a US Forest Service work crew, she became sensitive to several other chemicals which had never bothered her before. After she had recovered sufficiently from her initial exposure to work again, she got a job at a manufacturing plant with its own wells for drinking water. She became extremely ill within a few months with non-electrical seizures and asthma-like symptoms, and was forced to resign. At about the same time, it was discovered the plant well water had been contaminated with minute amounts of TCE. Most of the employees had not been affected, but, after extensive consultation with several therapists, it was determined that the patient had developed a full-blown case of Multiple Chemical Sensitivity, and was thus susceptible to effects from the minute amount of pollution.

In order to recover, she had to move into a small cabin deep in the mountains, far from any main road, with almost no artifacts of modern life present. Only after several years in this environment was she even able to tolerate short visits back to Denver, and then accompanied by an oxygen bottle.

HEAVY METALS

None of the heavy metals are a peculiar danger to EI or MCS patients. They are a general hazard as pollutants in themselves, and are all toxic.

LEAD

Symptoms of lead poisoning are wide-ranging. Mental problems and nervous system malfunctions are perhaps the best known, and are sometimes said to be a primary cause of decadent behavior during the later years of the Roman Empire, when lead was used extensively for water pipes. Children are especially vulnerable to the effects of lead, which can cause learning difficulties. Difficulty in absorbing nutrients is also associated with lead poisoning, and can lead to a host of other medical problems. Extreme poisoning can result in convulsions and death.

Up until the late 1960's, lead was a common component of interior house paints, especially the cheaper varieties. When lead poisoning was found among children living in slums of New York, a movement began to remove it from paint. Many children in those neighborhoods were habitually eating chips of old paint as it peeled from walls and baseboards. It has been noted in Prevention Magazine, circa 1978, that malnourished children are often attracted to heavy metal residues. Treatment with mineral supplements, especially zinc, has been shown to cause children to stop eating soil, paint chips, and other metallic residues. Personal experience in several situations has borne this out. Even feeding sunflower seeds, which are high in zinc, to young children has often helped.

Test kits for lead in paint are available through scientific supply houses and environmental specialty stores. They are generally easy to use. If you are about to repaint an older house, it is a good idea to test all layers of old paint. If lead is found in any layer, paint stripping will have to be carefully done while wearing breathing apparatus.

Lead has also been used as a fuel additive for many years, and thus was present anywhere near heavy traffic. Its phase-out in motor fuels in the United States and parts of Mexico has been well-documented in the press.

Water systems sometimes carry lead, making this a concern for many homeowners. Very old houses, built during an era when lead solder and sometimes lead piping was commonly used, are vulnerable. New houses, less than four years old, may sometimes show lead contamination, as trace amounts of solder leach into water. Pipe solder without lead is more difficult to work with, but is becoming more common. Fortunately, water lead testing kits are now widely available.

A common piece of advice, worth repeating here, is to run water for about two minutes in the morning before actually using it. This obviously applies to drinking and cooking water, and also showers, since lead can be absorbed through the skin.

MERCURY

Gold panners and processors have long used this element for separating gold particles from sands and ores. Therefore, for over a century it has been a critical water pollutant in parts of Colorado, Wyoming, Montana, and Nevada, among other places. In other parts of the country, it is used in manufacturing processes, and mercury poisoning has been well documented. In fact, mercury is among the first occupational hazards ever recognized, as evidenced by the "Mad Hatter" stereotypes that began appearing in literature in the 1500's. (Mercury was used in those days to cure furs for hats.)

Dentists have used mercury in amalgams for tooth crowns for a long time. Now, a great controversy is raging within the dental profession over the wisdom of removing old fillings. Mercury vapor levels up to 32 ppm have been measured inside some people's mouths! Some patients have reported feeling much better after the metal fillings were removed and replaced with newer, more neutral plastics. Others have reported that the activity seemed to release more toxins into the system, and they actually felt much worse for a time before getting better.

Most symptoms of mercury toxicity are mental, including memory loss, paranoia, manic episodes, and other forms of instability, and it can cause mutations, as demonstrated in the Minimata case of the early 1950's. Minimata was a Japanese harbor extensively polluted with mercury. It got into fish, which were eaten by young women who bore grotesquely mutated children.

Except for thermometers, thermostats, and barometers, mercury is not found much in ordinary households. Occasionally, it is an ingredient of fungicides and paints. It is dangerous if spilled, because it can migrate into cracks and then slowly vaporize. Even small amounts of the vapor can cause problems, especially in children. The effects are cumulative. So if a thermometer ever breaks in your home, clean it up quickly and thoroughly!

CADMIUM

This element and its salts are extensively used in alloys and paint pigments. It is not found much in ordinary household products. People who use acrylic and oil paints in their hobbies need to be careful. It is sometimes an air pollutant in neighborhoods near processing plants, as in the ASARCO case in Globeville, a neighborhood of Denver. The cleanup of cadmium residues there, mixed with other heavy metals, will cost up to fifty million dollars (figures vary depending on the source).

CHROMIUM, ZINC, MANGANESE

These three metals form a paradox. All three are used extensively in many manufacturing processes. They are also essential body nutrients in trace amounts. However, when found in the environment as particles, they are poisonous. Again, the main household problems with these stem from exposure through industrial air pollution. They can be found in some household products, usually automotive, but are relatively easy to control in the home. It is thus always a good idea to keep all automotive products out of living areas.

PENETRATING, POISONOUS SOLVENTS

"Halogenated" refers to any chemical containing one of the halogens, which are the second column in from the right in the Periodic Table of the Elements. These elements are all extremely reactive, and so rarely exist free in nature. Tremendous amounts of time and energy have to be used to make most of these substances. The halogens are, in order, Flourine, Chlorine, Bromine, Iodine, and Astatine.

Some of the most toxic substances ever made, including PCB's, CFC's, EDB, and many pesticides, fall into this large chemical family. Most chemicals in this class are by nature solvents, which means they will readily penetrate into the human body. Therefore, several of the halogenated hydrocarbons have been banned completely. Others have been banned in consumer products only, but are used industrially. However, residues of the chemicals, as used in manufacturing processes, can sometimes cling to products, thus causing problems for EI and MCS patients.

Almost any halogenated organic will be a problem for humans exposed to large amounts of the chemical. Most are listed as potentially toxic substances. Any of these could act as sensitizers, according to extrapolations from medical literature and interviews. As far as EI and MCS patients are concerned, they must rigorously avoid all of these, as they cannot personally afford to give any halogenated organic the benefit of a doubt.

One of the problems the human body encounters when trying to metabolize halogenated organics is the fact that Chlorine and Iodine are used in some metabolic processes. Chlorine is an essential part of the most abundant digestive enzyme, Hydrochloric Acid, which is itself an unstable chemical under some circumstances. Iodine is used in thyroid gland processes, and helps to regulate many enzymes related to growth, basal metabolic rate, and personal energy levels. Any halogenated organic could potentially interfere with either of these functions. As they break down in the body, numerous toxic metabolites (secondary compounds) are always formed. Some of the metabolites, being closer in chemical formula to normal enzymes, can end up being even more poisonous than the original substance. At times, a metabolite may be replacing a normal enzyme, and performing its function incorrectly. These reactions are usually too complex to track accurately, so some of the problems have to be extrapolated from the nature of systemic damage appearing in patients.

Following are brief summaries of the characteristics of a few of the most prominent halogenated chemicals. After that is a section on Trichloroethylene, followed by sections on other household hazards.

CARBON TETRACHLORIDE CCl_4

One of the simplest members of the halogenated organics family, Carbon Tetrachloride enjoyed a long term of popularity for many household cleaning activities. In 1970, it was banned for use in consumer products by the FDA because no warning label could possibly convey all its hazards.

Now, this chemical enters the home only incidentally, as a component of air pollution in industrialized areas, and clinging to materials that were processed with it (including dry cleaned clothes in some cases). Therefore, it is not a major problem in homes any more. EI and MCS patients usually

102

have to give up using dry cleaning services, in case this chemical or its relatives are being used.

One of the most important aspects of this chemical is its toxicology, because it is regarded as a model for all chlorinated organics. This is due to its long history of usage in many applications. Good case histories of problems with acute and chronic exposure date back as far as the 1920's. We know it can cause CNS effects, anemia, and liver damage. We also know that it creates several toxic metabolites, including chloroform, carbon monoxide, and phosgene, as the liver and kidneys attempt to deal with it. Most important to MCS sufferers, its interactions with other chemicals in the body, such as alcohol, acetone, chlordane, and phenobarbital, are well documented. In the future, we may be able to use old studies on Carbon Tetrachloride as a baseline of data on all chemicals in its class, assuming that it's all right to err on the side of caution when dealing with EI and MCS.

CHLOROFORM $CHCl_3$

Most people have heard of this chemical as an anesthetic. Its acute toxicity is fairly well-documented, causing the same anesthetic action that makes it so useful for surgery before going on to cause liver and kidney damage. MCS patients will almost always demonstrate a sensitivity to this chemical. Fortunately, few people encounter it in the home, except in some over-the-counter medicines, including cough medicines (notably some versions of Chloraseptic), cold sore remedies, and cold tablets. All these are easy enough to avoid.

1,1,1 TRICHLOROETHANE $CHCl_2CH_2Cl$

This deserves mention because of its presence in Scotchguard, a popular spray-on fabric protector. Also, dry cleaners and automotive cleaners use this chemical. It is considered "one of the least toxic of the liquid chlorinated hydrocarbons" (Gosselin et al. 1984). That is little comfort to someone who experiences severe reactions to treated clothing. Many EI and MCS patients have reportedly demonstrated sensitivity to this chemical, manifesting symptoms which include headaches, eye problems, and vertigo.

Mild skin and eye irritation have been reported upon contact. According to the NIOSH (US National Institute for Occupational Safety and Health) Pocket Guide, one of its alternative forms, 1,1,2 Trichloroethane, is a known carcinogen with an IDLH (Immediately Dangerous to Life and Health) level of 500 ppm. Liver and kidney effects are also listed.

It does have central nervous system effects, and is sometimes abused.

Also, interactions with other chemicals, such as ethanol and epinephrine (commonly found in cold remedies), have been reported. This means that exposure to 1,1,1 Trichloroethane may not be a problem under ordinary circumstances, but after exposure to other chemicals, serious reactions can develop. It is thus possible, because of a chance juxtaposition of two chemicals, that no one would ever find out exactly what caused a reaction.

EI and MCS patients must ask themselves whether or not they really need fabric protection. Considering their sensitivities, it's probably not worth it.

PENTACHLOROPHENOL (PCP, Penta) C_6Cl_5OH

Until the early 1980's, this was a common treatment and preservative for structural wood, railroad ties, and highway posts. It is still used in some engineering applications. Like many halogenated hydrocarbons, PCP is inimical to all life processes. The IDLH level is 150 mg/m3. That's what makes it such an effective wood preservative — no insect or microorganism could possibly attack anything impregnated with this substance.

So many people have had reactions to PCP that it has been largely withdrawn for consumer use in the United States. In the NIOSH Pocket Guide, there is an unusually long list of symptoms covering every body entry path. Its half life is 6 - 7 years, which means many structures still have vapors present. These vapors are easily detected by many EI patients. Although specific literature was not found, there is a strong possibility that PCP is a sensitizer. In some cases, PCP has been replaced by Formaldehyde, which can be almost as difficult to live with.

By the way, don't confuse this chemical with Phenylcyclidene, which is also abbreviated as PCP. The "other" PCP is a horse tranquilizer sometimes abused as a street drug.

In the Bau-Biologie material from Germany, borax preparations are often recommended as a substitute wood treatment, and then only when absolutely necessary. After all, throughout Europe there are wooden structures which have been standing for over 500 years with no wood treatment whatsoever.

Ordinary paints are usually sufficient to preserve exterior wood. Sometimes, oils are needed. In extreme environments, such as on seashores, wood preservation is absolutely necessary. In those cases, it is better to ask if the structures should be there in the first place (although hurricanes often provide Nature's reply to that question).

CHEMICAL FORMULA & DESCRIPTION C_2HCl_3

Trichloroethylene (TCE) is a halogenated hydrocarbon. It is thus in the same chemical family as Methyl Chloride, Methylene Chloride, Chloroform, Carbon Tetrachloride, and Ethyl Chloride. Many of its toxic effects are similar to these chemicals. TCE vapors form readily in many situations.

TLV: 50 ppm; Odor threshold 50 ppm; IDLH 1,000 ppm

USAGE & RELEVANT PROCESS

Light industry, dry cleaners, and auto repair shops all have uses for TCE. It is an excellent degreaser and solvent. Traces often remain in material after process steps. A few generally available cleaning fluids, especially for automotive uses, contain this chemical.

TCE is the agent of choice for the majority of decaffeination processes today. Several therapists have privately commented that they do not feel coffee processed in this manner is safe, because of trace amounts left in the beverage. However, research has not shown directly harmful effects from this process.

Since it is considered less harmful than Carbon Tetrachloride, it has replaced that chemical in some uses since the 1970 ban. OSHA lists it as a known carcinogen, and limits workplace exposure to 50 ppm.

Some wells near the Rocky Mountain Arsenal were contaminated with trace amounts of TCE, possibly because of industrial process wastes dumped on the grounds throughout the 1950's and 1960's.

INTRODUCTION TO THE BODY

Since TCE is classified as slightly water soluble, its vapors can irritate the middle respiratory tract.

The vapors have effects on the central nervous system and brain, so this chemical is sometimes abused. Symptoms of acute exposure include a type of intoxication.

Skin irritation on contact has been reported in the literature. It can penetrate the skin.

CHRONIC TOXICITY PROFILE

TCE attacks the liver and kidneys. Chronic, long-term exposures have definitely produced depression, alcohol intolerance, and increased cardiac output. Toxic centilobular necrosis (a kind of tissue death) has been found in liver biopsies of chronically exposed patients. Several metabolites, including phosgene, dichloroacetylene, and trichloroethanol can produce kidney failure and cancers.

Some EI patients have said their problems started or were intensified after exposure to TCE in water. This points to the possibility that TCE is a sensitizer, which is not noted in literature consulted for this section. In the late 1970's, some wells at a Denver-area manufacturing plant were contaminated with this chemical, and a few workers were treated for related symptoms, according to an informant.

SPECIFIC MITIGATION MEASURES

Most EI patients who have developed Multiple Chemical Sensitivity Syndrome need to avoid all dry-cleaned articles as a matter of course, in order to minimize exposure to residues of both TCE and Carbon Tetrachloride. As a precaution, many patients avoid drinking decaffeinated coffee, even in the absence of any conclusive evidence of harm from this commonly used beverage. To meet this demand, a few suppliers have come up with a water process that does not remove caffeine as thoroughly as TCE. Some people regularly buy bottled water in attempts to avoid this chemical in particular, since it can easily enter groundwater. Unfortunately, no legal standards exist for bottled water, so this can be a futile exercise. When in doubt, there are a few independent laboratories in the United States who will test water samples for this and other contaminants.

Most other exposures to this chemical have been occupational, so precautions have to be applied in the workplace.

CHEMICAL OUTLINE – AROMATICS

CHEMICAL FORMULA & DESCRIPTION
BENZENE C_6H_6
This is the first chemical discovered to have a ring-type structure. Its discovery through a dream of dancing snakes is a well-known scientific legend. From this came all the rest of the ring compounds, many of which are known as "Aromatics" because of their tendency to form odorous vapors.

TWA 0.1 ppm, IDLH 3,000 ppm
TOLUENE $C_6H_5CH_3$
Solvent effects are well-known among manufacturers and the general public.

TLV 100 ppm; Odor Threshold 0.17 ppm; IDLH 2,000 ppm

XYLENE $C_6H_4(CH_3)_2$

Three different forms, known as isomers, m-, p-, and o-, commonly exist in commerce. Although they have widely varying flash points and flammable ranges, their effects on the body are almost identical.

All isomers: TLV 100 ppm; Odor Threshold 0.05 ppm; IDLH 10,000 ppm

OTHERS

Also in this chemical group, but rarely seen, especially in the home, are Cumene ($C_6H_5CH(CH_3)_2$) and Mesitylene (formula not found). For purposes of toxicity studies they are sometimes placed with Xylene, and treatments are the same. A related chemical, Toluene Diisocyanate (TDI), is extremely toxic and can cause serious respiratory disease, but fortunately is used only in industrial processes. There are many other chemicals in the Aromatic family, and also many derivatives of these substances used throughout the world. All of them should be rigorously avoided by EI and MCS patients.

USAGE & RELEVANT PROCESS

BENZENE

Long used as an industrial solvent, basically since the day after its discovery in the 1800's. By the 1930's, reports of anemia were beginning to surface among people working with it. Eventually, it was discovered that Benzene directly attacks the bone marrow, which produces red blood cells, and it was gradually withdrawn from many uses. It was banned by the Consumer Product Safety Commission for use in household products in 1978. By 1979, Benzene was firmly established through animal studies as a carcinogen. It is still present in cigarette smoke, some paint vapors, and a few automotive products.

TOLUENE

As Benzene became less accepted for industrial uses, Toluene took over many of its functions. It also became the base for many types of glues and cements, especially those intended for plastic model hobbyists. It has been used in many paint formulations, cleaners, and insecticides. Since that time, in the mid-1950's, abuse of this chemical for psychotropic effects (glue sniffing) has become intermittently fashionable in many parts of the world.

XYLENE

Manufacturers noticed that Xylene was a little easier to handle than Toluene, and so it has, in recent years, become a more popular solvent. In its pure form, it is most often found in furniture stripping compounds. Other uses are exactly the same as for Toluene. It is not considered as hazardous as Toluene, mainly because of its higher IDLH value. Several studies have indicated that Xylene is similar to Benzene in the way it is metabolized in the body.

INTRODUCTION TO THE BODY

All three of these chemicals produce vapors which can easily be inhaled. Benzene and Toluene are irritating to the skin and eyes. In general, Xylene is not as irritating. All three of these chemicals can cause liver and kidney damage. Serious and fatal poisonings have been known to occur solely through inhalation of high vapor concentrations. Lower concentrations can produce wide-ranging symptoms of intoxication.

Skin absorption can also cause problems, usually associated with destruction of protective fat.

Ingestion is of course a possibility, but it doesn't happen often. Fortunately, small children tend to dislike the smell of these chemicals and usually leave them alone.

CHRONIC TOXICITY PROFILE

Anemia is the classic end result of long-term exposure to Benzene. Many otherwise rare blood conditions have also been linked to Benzene, as have blood and bone marrow cancers. It causes surprisingly little liver and kidney damage.

Toluene causes some blood cell changes, which can be accumulative if exposure persists for too long or too often. It has been known to produce severe muscle weakness, mental deterioration, and psychosis.

Xylene may accumulate in fatty tissues, even after being inhaled. This has a lot of long-term implications, and means Xylene can be a sensitizer. There is an absence of solid data on chronic effects of this chemical, although some sources state its effects are simply "similar to Toluene".

SPECIFIC MITIGATION MEASURES

All three of these chemicals can be avoided in the home, mainly by curtailing hobbies where glues, paints, and furniture strippers are commonly used. Do-it-yourselfers who have demonstrated sensitivity to these chemicals can fall back on older carpentry methods, such as use of linseed or other vegetable oils to preserve wood, careful sanding to strip old layers of paint, and animal product glues. Choosing paints without the aromatics is difficult, but it can be done. Read labels carefully when shopping! The average hardware store clerk rarely knows anything about toxic chemicals. Usually manufacturers have to be consulted directly to obtain good information on this subject.

CHEMICAL OUTLINE – ACETONE

CHEMICAL FORMULA & DESCRIPTION CH_3COCH_3

Also known as propanone, and sometimes dimethyl ketone. This is one of the more volatile organic compounds. As an example of its volatility, in a demonstration, a rag soaked in Acetone burned completely, with no ash left over, in about three seconds.

TLV 750 ppm; Odor Threshold 100 ppm; IDLH 20,000 ppm

USAGE & RELEVANT PROCESS

In the home, Acetone is most often found in nail polish removers. Acetone is in many kinds of glues, since it is an excellent solvent and evaporates quickly. It is also used in printer's inks, and in a few remedies for burns and calluses. Acetone is the preferred buffer compound for acetylene, and so is widely found in welding shops, as well as in manufacturing processes worldwide as a solvent or component. It is commonly used in shipping as an inhibitor to keep components of plastics from exploding.

INTRODUCTION TO THE BODY

When inhaled, Acetone quickly anesthetizes the sense of smell. It is thus easy for someone to be overdosed without knowing it. They will think it has gone away, while it is still present in the air. It can make a person act drunk.

During acute exposure, skin rashes usually develop.

If ingested, Acetone causes a great number of symptoms, and can be fatal in large quantities.

CHRONIC TOXICITY PROFILE

Acetone is an irritant, especially to chemically sensitive people. It is known to cause headaches and skin rashes. No cases of a role as a sensitizer have been found. Its effects are apparently worst in people who already have multiple chemical sensitivities.

SPECIFIC MITIGATION MEASURES

Acetone is usually easy to avoid in the home. Its use as a nail polish remover is questionable, because it is known to make nails brittle if used too heavily or too often. Besides, many men have long questioned the attractiveness of polished nails. Glues and over-the-counter remedies containing Acetone are replaceable with substitute milk-based products. If someone develops sensitivity to Acetone, it is likely that many other chemicals will also cause adverse reactions, and Acetone will become just one of many substances to be avoided.

CHEMICAL OUTLINE – FORMALDEHYDE

CHEMICAL FORMULA & DESCRIPTION   HCHO

Formaldehyde is a basic laboratory preservative and reagent. It is related to Formic Acid, which is a natural compound secreted by ants, bees and wasps in their stings. Its structural simplicity causes it to penetrate almost anywhere, and stick to fibers. Other names include Formalin and Formol.

TLV: 2 ppm; Odor Threshold 0.8 ppm; IDLH level 100 ppm.

USAGE & RELEVANT PROCESS

Formaldehyde is one of the most common household chemicals. Generally, it is not directly used in household products, but instead is an incidental component of materials brought into the home due to manufacturing processes. This includes curtains, clothing, plastics, bedding, and even paper. It is usually present in new pressboard furniture. Formaldehyde has replaced PCP as a wood preservative in many applications in the United States and Europe. It is a component of Urea-Formaldehyde insulation, which was sprayed into a lot of houses and mobile homes between 1973 and 1982 in pursuit of energy efficiency, until a lot of reports of reactions piled up. Urea-Formaldehyde insulation was banned for residential use in 1982. It is commonly used with monomers in plastics manufacture, and is an essential component of Phenolics and Aminos.

Among the household products containing formaldehyde are some glues and mildew-proofing compounds. Sometimes, it is in shampoos under the alias "Quaternium-15".

Formaldehyde remains the most used embalming fluid, as it has for many decades.

Several close relatives of Formaldehyde sometimes are found in households. These include Acrolein, which is much more of an irritant than

Formaldehyde, and is sometimes made by overheating cooking oils and animal fats. Paraformaldehyde is found in a few deodorants and causes allergic reactions in some people. Sometimes Glutaraldehyde appears as a pollutant in wood and cigarette smoke. In fact, all the aldehydes are present in most household smokes.

INTRODUCTION TO THE BODY

EI patients have reported problems through both skin and respiratory pathways. In fact, according to several people interviewed for this section, this is, in their view, the most insidious of all irritants, because of its extensive use in building materials and because visitors to a home may inadvertently bring it in on new permanent-press clothing. Many patients routinely ask all friends to be especially careful of the clothes they wear when visiting, usually requesting only well-worn cotton or linen. This is no guarantee of safety, as reports of formaldehyde persistence have surfaced even after several washings.

Death can occur through ingestion of about 1 ounce. It would be a particularly nasty way to go.

CHRONIC TOXICITY PROFILE

Dermatitis and severe coughing are the two most common symptoms of chronic toxicity. Some patients have also reported eye problems such as persistent conjunctivitis-like symptoms after any exposure, which has not been noted in the literature.

In extreme cases, patients have reported severe, debilitating headaches, constant flu-like symptoms, and chronic fatigue. In these cases, it is surprising how little of the chemical residue can bring on symptoms — often amounts that are almost impossible to measure. Only the human body is sensitive enough to measure some things.

SPECIFIC MITIGATION MEASURES

Learning all of Formaldehyde's current aliases and relatives is an essential component of survival for MCS patients. Avoiding this chemical can mean eliminating much of modern civilization from the home. A few catalogs now offer clothing processed without Formaldehyde, and it is relatively easy to avoid glues, soaps and shampoos containing it. Curtains, carpets, building materials, and bedding without it are much harder to find at this time. A few makers of futons are offering products processed without it, probably because much of their existing market is to people who have

111

environmental concerns.

Many people who are avoiding Formaldehyde remove all carpets from their homes, and live on hardwood floors. They must diligently seek out the specialized catalogs and stores which offer chemical-free products.

CHEMICAL OUTLINE – PHENOL

CHEMICAL FORMULA & DESCRIPTION C_6H_5OH

Also known as Carbolic Acid, this chemical has a relatively simple ring-type structure.

TLV: 5 ppm; Odor threshold 0.5 ppm; IDLH 100 ppm

USAGE & RELEVANT PROCESS

Its simplicity compared to some other ring compounds makes it a basic monomer, used extensively in plastic production. Biologically, it is opposed to almost all life processes, which made it useful as a disinfectant before its dangers were known. One variant of Lysol still contains a phenol derivative, according to Consumer Reports, January 1988. Some industrial cleaners have phenol in the formula. Winter notes that it is also used in some shaving creams and hand lotions. Clinical Toxicology of Commercial Products noted its presence in several over-the-counter medicines.

According to one moderately environmentally-sensitive person, Phenol was often sprayed on ponds during the 1960's to kill mosquito larvae, and she attributed several current health problems to exposure during her youth in Longmont thirty years before.

INTRODUCTION TO THE BODY

Acute toxicity is well documented, with a fatal dose by ingestion established at about 1.5 grams. Ingestion of smaller amounts can cause serious digestive tract disturbances, with eventual circulatory system collapse. Toxicity through skin absorption is possible, with many cases noted.

CHRONIC TOXICITY PROFILE

Phenol is regarded by many EI patients as one of the single most dangerous chemicals known. Its name came up many times in interviews. Some patients think that it remains on some plastics as a trace residue, and thus can trigger reactions. This has not been well-documented.

Reports have surfaced saying that phenol has terrible long-term effects. Rogers regards it as one of the worst environmental toxins, often triggering

112

reactions for many years after exposure. She feels it is a sensitizer, and has listed almost every EI symptom next to Phenol at one time or another.

SPECIFIC MITIGATION MEASURES

Environmentally sensitive people must avoid Phenol at all costs, by reading ingredient labels. Since most exposures in these patients cause immediate skin reactions, discontinuance of products with no ingredient listings is considered mandatory at the first sign of a problem. In the most extreme cases, plastics in general have to be avoided, because of the small possibility of Phenol residues. Fortunately, many authorities have already discontinued its use in public health and medical applications, with a few exceptions.

One problem with Phenol reported by Rogers (herself reporting on work by other allergists) is that this chemical is often used as a base for allergy extracts. These extracts are extensively used in treatments, and may themselves cause allergic reactions in some patients because of the Phenol. It is possible to obtain Phenol-free extracts, although they are more expensive.

RADIOACTIVE COMPOUNDS

CHEMICAL FORMULA & DESCRIPTION

Radon gas (Rn) and other radioisotopes

Decay process daughters, too numerous to mention, which usually exist for short times as isolated units constituting just a few atoms.

USAGE & RELEVANT PROCESS

Radioisotopes are fortunately rare in most homes. The general public is well aware of radon gas by now, since reports on problems have occurred regularly since the late 1970's. The level of alarm among the public seems far in excess of the actual danger in most cases. In fact, many people immediately associate environmental inspections with radon, without even a thought about more urgent and dangerous problems.

Radon generally seeps into a house through contaminated soil, and then begins to decay. The contaminated soil is worst when it involves uranium mining and process waste. In Grand Junction, Colorado, uranium tailings were used as fill for many homes, commercial buildings, and even parks. The cleanup job there is tremendous, and has even necessitated the building of a special water pretreatment plant. Actually, tailings from any kind of mining operation are liable to contain radon gas. Radon does not ever exist in nature in any great quantity, being itself a product of other radioactive elements as they decay. When ground has been intensively disturbed by mining, the natural decay processes may be disrupted and many elements will be concentrated into the tailings, thus making them dangerous. Radon contamination might occur anywhere, but it is most common in the United States in the Rocky Mountain region and in Florida.

Although Americium (and sometimes Californium) are often used in smoke detectors, thus making the presence of at least one radioisotope an ordinary fact of life in many homes, the amounts present are so negligible that only a few researchers have called it a danger. People experimenting with Orgone energy, as described by Wilhelm Reich, should be especially careful. Peter Lindemann has recommended that anyone conducting Orgone experiments must first remove all smoke detectors and any other radioactive emitters. Nonradioactive smoke detectors are now available, usually based on photoionization.

INTRODUCTION TO THE BODY

Under normal circumstances, nobody is exposed to radiation of sufficient strength to cause skin burns or other dramatic symptoms. This can only happen during the most serious accidents or from nuclear detonations. Still, people can get very frightened of radiation, because it can't be seen or smelled, and is relatively difficult to detect.

Radioactive materials are most dangerous when inhaled. This can potentially occur with radon gas and some of its daughters, which are isolated atoms occasionally floating in air. Radon is slightly heavier than air, so it usually tends to stay near the ground. Therefore, people sleeping in the basement of a contaminated home would be in the most danger.

When atoms decay inside the body, there is a release of particles which can damage tissue or cause cellular mutations. Radon and its daughters usually emit alpha particles, which are isolated helium nuclei. These nuclei are actively seeking electrons to complete the atom. It is possible that the necessary electrons could be "stolen" from metabolic processes, thus causing some weakening of the body. Alpha particles can also directly cause cell mutations by physical action, which can eventually evolve into cancers. The end of radon's decay cycle is always Lead, which is itself highly poisonous.

114

CHRONIC TOXICITY PROFILE

Cancer is the most common end result of low-level radioactive contamination. Where radon gas is involved, lung cancers are most common. Still, cancers only occur in a small percentage of cases where radon has been present in sufficient quantities to be termed dangerous. It is usually more a statistical probability than a direct threat. Smokers are always much more susceptible to cancer than everyone else when exposed to high amounts of radon.

SPECIFIC MITIGATION MEASURES

Testing for Radon is relatively easy. Canister kits are now widely available at hardware stores, and are normally utilized by simply unsealing and leaving in a vulnerable, unventilated area for about three weeks. They are then mailed to a laboratory for analysis. Spot checks do little good in finding Radon, because its release and decay are random events occurring over a period of time. (Most affordable geiger counters can't measure alpha particles very well anyway, and Radon is primarily an alpha emitter.)

Some building inspectors use complicated, expensive equipment to test for Radon. This may be necessary in some situations, where people might come in and disturb ordinary canisters. Actually, the equipment is not any more accurate than the cheap canisters. Unfortunately, Radon tests are some of the easiest to skew by doing improperly. The area being tested MUST be kept unventilated throughout the whole test period. Some homeowners have been known to falsify tests by leaving all windows and doors open in order to get a good test in a house sale. Obviously, tests are harder to conduct in the summer, so tests in North America and Europe are best done from November through February.

The U.S. Environmental Protection Agency has recommended specific action levels, based on test results, to mitigate radon gas in homes. These action levels should in most cases trigger a remodelling program designed to increase outside air intake for the home. Depending on levels found, the lower floor of a house may need to be specially ventilated and isolated from the rest of the house. Sometimes, simply cutting vent holes into a basement or crawl space is sufficient. In extreme cases, the EPA has recommended active negative pressure systems to constantly draw in fresh outside air. These systems can be expensive, difficult to maintain, and noisy. They may also be the only choice a homeowner has in order to salvage equity in a heavily affected home.

Some building materials can be toxic, because they tend to contain certain chemicals. This can be a difficult and technical subject. Anyone doing remodelling or new building would do well to consult an environmental expert before ordering materials. Since manufacturers change processes, and architects favor different materials from year to year, updates are happening constantly. Testing each proposed material for toxins is, of course, prohibitively expensive. Keeping in mind some of the chemical names mentioned earlier in this section, always read labels when buying materials, especially caulk, insulation, and paint. Here are a few materials which might cause problems under certain conditions.

LATEX: A type of artificial rubber often used to give a good hide to paint. Most latex-based paints are water-soluble and at least somewhat biodegradable. There is some outgassing from latex paints for a few days after application. A few people are allergic to any form of latex and its vapors.

MINERAL WOOL: A good insulator, this was popular in parts of America for many years. It is good when contained, as it does not tend to outgas, but if it is disturbed, the dust problems can be horrendous. Its dust is irritating to the skin and respiratory tract. This is one good reason to always wear a dust mask when going up into an attic. There have been cases of this material getting into attic furnace flue systems, which can be a nightmare.

DRYWALL: There are big differences in drywall composition between America and Europe. A lot of American drywall is made of compressed gypsum, and is relatively stable unless it gets wet. Some types are excellent fire retardants. In newer developments, drywall is everywhere, and we are fortunate that the material itself tends to be neutral. The finish, however, can include many noxious chemicals. Some drywalls, especially in Europe, have a high content of artificial ingredients, which can outgas and cause health problems.

POLYSTYRENE: Sometimes used to insulate foundations and walls. There may be some outgassing in poorly manufactured material. Its manufacture causes a large amount of pollution, and it is persistent in the environment. Present estimates say that most types of polystyrene particles, after becoming useless, will still persist for 1,000 years before breaking down. Even if the stuff doesn't hurt you as an individual, it's hurting everyone indirectly just by being made.

POLYURETHANE: A foam used in insulation and sometimes caulk.

It can expand to fill space, which makes it popular for remodelling. Unfortunately, it creates toxic vapors which some people react to violently. It has been banned for certain uses in the United States since 1982.

VENEER BOARD: Many veneer boards contain or require glues which can outgas.

BITUMINOUS FELT: Often used as a roofing material, this is acceptable if it is sufficiently isolated from living areas. Some types of bituminous felt will give off odors when heated.

ASPHALT ESTRICH: Anything with asphalt arouses suspicion, because it can too easily give off noxious odors when heated.

SOFT WOODFIBER BOARD: When used as a jacketing, it can have some insulation value. Again, there can be problems with chemical outgassing, although it is not severe with most brands. This type of board is easily attacked by moisture, which means that vapor barriers may be required with it. This creates other difficulties, especially if the vapor barriers are improperly installed.

CARPETING: As of this writing, huge controversies are raging within the industry because of several EI cases apparently triggered by poor installation. Much of the debate centers around glues and a compound added to carpet fibers known as 4,PC. At this time, nobody honestly knows whether or not 4,PC is really dangerous. It is better to avoid carpeting altogether until the situation is resolved. Use natural-fiber area rugs instead.

TREATMENTS FOR ENVIRONMENTAL ILLNESS AND MULTIPLE CHEMICAL SENSITIVITY

It has been extraordinarily difficult to treat Environmental Illness (EI) and Multiple Chemical Sensitivity Syndrome (MCS). Two primary factors are responsible for these difficulties. One of those factors is the political and social structure of the American medical system, which will not be covered in this summary. Suffice it to say that because of this structure, meaningful research, tabulation of results, and communication between therapeutic disciplines have been hindered ever since the discovery of these two problems.

The nature of the syndromes is the other factor making treatment difficult. For many patients, the onset of symptoms is gradual. The exact symptoms will vary widely from one case to another. Sometimes unexplained, spontaneous remissions happen. However, many cases are allegedly a grim spiral down to progressively worse and more difficult symptoms, until the patient either dies, changes doctors, or begins a strict course of avoidance and dietary regimens. Jumping from one therapist to another is common among EI and MCS patients. Many of these people simply have to move as far out in the country as possible.

Practitioners in several different disciplines have made efforts to treat these syndromes. Following is a brief summary of a few methodologies currently practiced.

ALLERGY MEDICINE (CLINICAL ECOLOGY)

This approach is well-documented in the books by Sherry Rogers. Diagnoses are typically made by skin patch tests, some of which can themselves be devastating to a patient. These methods can be crude in some ways, but they are often the only ones available through conventional medical channels, under insurance codings, at this time.

Treatment is a combination of avoidance, dietary restrictions, and re-testing. Each program has to be strictly individualized. The dietary restrictions often take the form of complicated "rotation diets", where certain classes of foods are eaten for a few days at a time, and then substituted with other classes for another few days. The possibilities inherent in rotation diet systems are infinite. They have proven effective for many patients, according to interviews.

Clinical Ecology, as presently practiced, is basically a subset of allergy practice. This is necessary in most states because of insurance company codings, which have never been set up to deal with EI and MCS. Clinical Ecologists tend to be so busy that it was impossible to set up an interview with one.

NUTRITIONAL THERAPY

Generally, therapists in "alternative" disciplines will take this approach, usually in combination with the therapist's specialty. The key is organic foods. No trace, even imagined, of pesticide residues or drugs can ever cross the patient's table. Eating in most restaurants is strictly forbidden, as the risks are too high. Partially or fully vegetarian diets are sometimes recommended, either because of suspicions about America's meat supply, or because of a disciplinary bias.

Any food considered to be extremely mucous producing is generally to be avoided, on the assumption that excess mucous will tend to obstruct elimination of toxic residues and metabolites. In most nutritional approaches to EI and MCS treatment, dairy products are considered the worst

118

possible foods. Fruits, whole grains, and vegetables with a lot of bulk are usually recommended.

Nutritional supplementation may be recommended. Usually, B vitamins are the most important, as they are said to assist in clearing toxins from the body. Carefully chosen mineral supplements, which must take into account the patient's chemical exposure history, may also be used.

CHIROPRACTIC TREATMENT

Most chiropractors who have graduated from school within the past ten years have had extensive nutritional training. They tend to stress this, while giving adjustments to spinal areas and specific muscle groups on a weekly or monthly basis. Since 1980, kinesiology has become more popular within the chiropractic community. Usually, weaknesses in specific organs can be identified rapidly through this technique. Chiropractors who deal with EI and MCS patients generally rely a lot on kinesiology, because internal organs are most affected, and muscle groups may not be affected. There are many chiropractors emerging within the past few years who rarely perform conventional spinal adjustments, instead preferring subtle kinesiology manipulations of the hands and feet, sometimes combined with meridian activation using acupuncture or magnets.

ACUPUNCTURE

Promising results have occurred in many EI and MCS cases, according to patient interviews. While basic principles of this practice are still considered obscure by most American medical personnel, we cannot argue with results. Specific body meridians are targeted with needles, which are used to alternately open up and block subtle energy flows to specific organs. When Chinese herbs are combined with a regular acupuncture program, effectiveness may be increased.

Patients are sometimes encouraged to use finger pressure on specific points when encountering difficult situations outside the clinic. A good example of this is a point close to the origin of the thumb, approximately 1/2 inch from where the bone forms the first joint by the palm of the hand. Pressure on this point tends to "quiet down" the liver, if there is indigestion (or a harmonic symptom, such as an itch in the right eye) present.

Officially, few in the American medical profession will say why acupuncture and acupressure should work. To consider this information seriously would contradict many fundamental principles of medical practice as currently understood.

HOMEOPATHY

Not enough homeopaths are successfully practicing in America to give this discipline a fair evaluation. European informants have explicitly stated that homeopathy is extremely effective for treating EI and MCS. Their comments indicated that treatment will usually be more intense, meaning more frequent consultations, when these syndromes are involved. Almost all homeopaths, when confronted with an EI or MCS case, will recommend strict adherence to an organic-food diet and avoidance of several aromatic foods, including mints, garlic, and onions. It is said that these foods will interfere with the subtle action of the homeopathic preparations. European homeopathic protocols may not always work when directly transferred to America, because the types of household chemicals used there are different.

HERBAL MEDICINE

Due to the lack of practitioners, lack of research, and difficulty with authorities such as the FDA, it is impossible to give a coherent and authoritative evaluation of herbal medicine programs for EI and MCS patients. Several herbs are said to clear toxins from the body. Foremost among these is Chaparral, which has an incredibly bitter, obnoxious taste. (Chaparral can be dangerous when used carelessly.) Others recommended for clearing toxins from EI and MCS patients have included Hyssop, Sassafras, and Sarsaparilla. Oregon Grape leaf is generally agreed on as the most effective herb for balancing and strengthening a damaged liver. Most herbalists also recommend Ginseng, Dong Quai, or Fo Ti Tieng as part of a program to build up general metabolic strength. As a specific for clearing out radioactive contamination, Sodium Alginate has often been mentioned, mostly in underground literature.

Any herbal program is of necessity going to be extremely individualized. Since good herbal practitioners are rare, most patients who have experimented with this idea have been doing so based on literature searches, networking underground with similarly inclined patients, and personal experience. Fortunately, the action of most herbs is gentle enough that a mistake can be easily corrected. There are a few combinations, however, which can be deadly, so caution is advised when practicing without supervision. An excellent herbal resource is a CD-ROM called The Herbalist, authored by David Hoffman and available through Hopkins Technology.

FINDING ENVIRONMENTAL TOXINS IN THE HOME

Testing for environmental toxins in residences is very different from what is necessary for industrial and toxic waste situations. In many cases, the amount of chemicals causing a problem is below detection limits for instrumentation and even standard laboratory tests. Sampling procedures sometimes have to be extensively modified, partly because much of the equipment was designed for use in industrial environments.

Another problem with residential chemical testing is, frankly, there is not a whole lot of money in it at this point. Few homeowners are in a position to spend thousands of dollars for specific lab tests. Most residences are extremely complex, consisting of several areas in which toxic contamination can take radically different forms. Developing site characterizations using the same protocols necessary for conventional EPA-supervised toxic waste sites would be prohibitively expensive.

Public awareness creates more setbacks for professionals in this field. The pioneers who are currently doing residential chemical surveys find that most people think there is some instrument out there which can detect every possible toxin in a matter of minutes, and will fit in a briefcase. This is, of course, not the case. Instrumentation is very expensive, and, in most cases, quite limited in its capabilities. Therefore, even residential specialists have to rely on laboratory testing and use good sampling procedures. An experienced residential environmental consultant must rely heavily on the sense of smell. Based on experience, intuition, knowledge, and careful observation, the consultant must then carefully choose a small number of probable pollutants, and order tests for those.

INSTRUMENTATION

Only a very few of the myriad environmental testing instruments are worth using in homes. Of course, experiments are going on all the time with available and new products, to see if they will work or can be adapted. So far, most toxic gas indicators have been eliminated, because they are designed to find only IDLH concentrations. Their sensors cannot easily be modified to accommodate trace amounts of the wide variety of chemicals typically found in homes.

Photoionization (PID) and Flame Ionization (FID) detector instruments generally have detection limits too high to be useful in residences. Furthermore, even more sophisticated equipment, such as the Foxpro 3000, provides at best a good guess, based on computerized calculations derived from a special calibration gas.

The new portable Gas Chromatographs might be useful for residences. However, they are currently too expensive to be practical as an addition to a residential specialist's arsenal.

One good possibility for residential work is a simple broad-band gas detector, such as the TIF 8800. These can give a consultant a good outline of a highly toxic area in a home, if one exists. Then, sampling and detailed lab analysis can be carried out from there.

AIR SAMPLING

Residential air sampling has as its goal the provision of good laboratory samples for further analysis. It is not practical or affordable to sample every bit of air in a house. Therefore, a consultant needs to target areas based on a combination of observed usage patterns, reported symptoms, history provided by the homeowner or real estate professional, and an initial survey. Most consultants like to work with therapists to get coherent case histories when possible, but that does not happen often enough at this time.

Once the target areas are determined, several pieces of sampling equipment can be used. Personal air samplers designed for hazardous industrial environments are easily adapted for residential surveys in cases where a specific chemical is targeted and a low flow rate will produce a good sample. Unfortunately, much of the available calibration equipment is fragile, which has been a severe handicap for consultants. Larger high-flow sampling pumps, such as the Andersen units, are useful for mold cultures and sampling for trace chemicals. Sometimes, special sample collection bags can be used.

Customized sampling media have recently become available for form- aldehyde, ozone, and carbon dioxide. Some of these media are colorimetric tubes available through Draeger and Sensidyne. Air Quality Research of Berkeley, California has formaldehyde sampling vials available. These are all designed to test for the low concentrations of chemicals typically found in homes, and so can be useful to residential specialists.

SURFACE AND WATER SAMPLING

Swab sampling kits for Lead are now widely available to homeowners, through hardware stores. These are adequate for estimation, but care must be taken to ensure that all paint layers are accounted for. Lead content may vary widely in the space of even a few inches, which can skew sampling results one way or another.

Test kits for Lead in water are widely available. If the test procedure is done correctly, a homeowner is in a position to get a fairly accurate reading.

Wipe samples for specific chemicals such as PCB's, Chlordane, and aromatics, are easy to take in residential applications, as the techniques are identical to those for industrial and hazardous waste operations. Since most homeowners are unaware of the availability of these more specific tests, it is the job of a residential environmental specialist to find a source and laboratory for these tests and arrange them for the homeowner if needed.

LABORATORY RESOURCES

A few laboratories around the United States are becoming more sensitive to residential survey needs. Any laboratory willing to bend a little for residential customers deserves a commendation, because at the moment they cannot make much money on this compared to what can be obtained from consistent contracts with large hazardous waste consultants and industrial customers. Still, it is best for residential surveys to be coordinated by a specialist who can work with labs and often arrange more favorable rates.

Deserving of special mention in this regard is P & K Microbiology Services of Cherry Hill, New Jersey. One of their microbiologists, Dr. Chin S. Yang, has written extensive criteria for residential sampling of aerosols, including excellent notes on sample integrity, safety, and general procedures.

CONCLUSION

Residential environmental testing is in its infancy in America. As of 1993, only a few specialists had been in business for more than three years. At this point, most practitioners are looking constantly towards Europe (and, in some cases, Taiwan) for techniques and advice. European consultants have generally been most concerned with electromagnetism and geobiology, and have not extensively developed residential chemical analysis techniques. Laboratory tests there tend to be approximately 100 - 200% more expensive than in the United States. A few researchers are beginning to try out Frequency Spectrometers, but those techniques and protocols are strictly experimental at this time. Taiwanese (and other overseas Chinese) consultants are primarily concerned with ancient protocols developed within the context of Feng Shui, and are only now getting into chemical and electromagnetic analysis. Therefore, much of what is happening in terms of residential environmental surveying in the United States is a learn-as-you-go

123

proposition, and new developments are occurring daily.

HANDLING HOUSEHOLD ENVIRONMENTAL HAZARDS

Avoidance is the key word, found repeatedly, whenever Environmental Illness (EI) or Multiple Chemical Sensitivity (MCS) is a factor. Specific chemicals can be avoided by careful label reading, backed up by knowledge of specific aliases for certain chemicals such as formaldehyde. In many cases, this pattern of avoidance extends to moving into another home, in an area as unpolluted as possible, to assist in recovery. On the other hand, it is sometimes possible to rehabilitate a contaminated home.

REMOVAL OF CONTAMINATION

Once a house has been contaminated, removal of toxic chemicals can be difficult. Wall-to-wall carpeting usually has to be removed altogether in these cases. Any contaminated spots on the floor or subfloor must be scrubbed thoroughly with a nontoxic detergent. If an old wood floor is present, that's good, because natural-fiber area rugs can simply be thrown on top. In more modern houses, where the subfloor was specifically designed for wall-to-wall carpet, Bau-Biologists often recommend going to the trouble and expense of laying an untreated wood parquet floor over the subfloor, once contaminated spots have been addressed. Parts of the subfloor may have to be replaced in extreme cases. (Ceramic tiles are also chemically neutral, but they create problems with heating.)

Sometimes, heating and air conditioning systems get contaminated with chemical residue. There have been many cases, in new construction, where solvents, old paint cans, and other trash were simply left behind in an air plenum. In cases like this, the chamber will have to be disassembled and cleaned. The ductwork will have to be cleaned thoroughly with a mild detergent. If the ductwork is of the fiberglass type, with a porous surface, it might have to be replaced.

If paint or walls are contaminated, especially when lead-based paint is present, most authorities recommend thorough sanding and HEPA-filter vacuuming. All workers involved must wear appropriate Personal Protective Equipment, including respirators and coveralls. Residents of such a house must move out until this process is complete.

MANAGEMENT OF EI AND MCS CASES

Most EI and MCS patients have to make major adjustments in their lives.

For example, clothes are never going to come out as clean with alternative detergents as they did with standard products. Housecleaning will take more elbow grease. Visits to museums, concerts, and theaters can be a nightmare, because one never knows when someone wearing too much perfume will come along. That's why many choose to simply leave populated areas altogether, and live in virtual isolation.

Adjustments are made more difficult because many symptoms of EI and MCS are mental, producing irritability, altered perception, and heavy emotional swings. Administration of psychiatric drugs generally makes the problems even worse, and will multiply case management difficulties. Even though a patient may be improving, tremendous strain is imposed on everyone living around an affected person, especially if they are not similarly affected. That said, occasionally people have hidden themselves behind a diagnosis as a convenient way to control others and escape responsibility for their own actions. All this must be taken into account when dealing with EI and MCS patients. It takes a strong, intelligent, and resilient person to live around someone suffering from these syndromes.

Professor A. Delgado of Metro State College in Denver built a house for his environmentally ill wife in a new subdivision between Castle Rock and Franktown. A few other EI and MCS patients have set up in that area as well, forming a small community. Several similar communities have sprung up around the country, including ones near Yuma, Arizona, and in central Texas. Assuming money is available for new housing, this is one viable alternative for case management, because the patients tend to be excellent resources for helping each other. Formal and informal communications are constantly going on between patients about sources for uncontaminated household items and alternative remedies.

A few catalog and delivery services cater to EI and MCS patients. Among those are two in Colorado, Allergy Resources of Palmer Lake, and An Ounce of Prevention inside Denver. There are some national catalogs, such as Seventh Generation, with sections including clothing prepared without chemicals, and alternative cleaning products.

PREVENTION

Prevention of EI and MCS is a difficult but worthwhile pursuit. It is impossible to know in advance who will become sensitive to small amounts of chemicals described earlier, or who will be accidentally exposed to pesticides. Several rules of thumb come to mind.

First and foremost, is the avoidance of harsh pesticides and herbicides.

Even currently approved, "gentle" pesticides have been said to cause sensitization in some cases. More intensive hygienic practices in the home, such as more frequent cleaning, and use of old home remedies such as Pennyroyal and Boron solutions will have to do. Constant research is ongoing into biological pest control methods, and in some cases these new methods have worked well.

Contractors coming to a house for remodeling need to be carefully monitored. Their trash disposal methods should especially be noted. Although most contractors tend to resent the interference, it is a good idea to question the use of any chemical on site that seems inappropriate at the time. Carpet and vinyl flooring layers can be the worst offenders for introducing toxic substances to homes. Because of this, a few smart contractors are changing their practices, and it is worthwhile to seek them out. Also, retaining a Design Ecology, Bau-Biologie, or similar environmental consultant can be an excellent way to help keep a contractor in line.

It is difficult, but not impossible, to purchase carpeting free of formaldehyde and other chemical residues. Natural fiber carpeting is available nationwide through an increasing number of suppliers. One promising alternative is carpet made from recycled plastic. The assumption is that the plastic long ago outgassed any toxic chemicals in its previous incarnation as a pop bottle. Unfortunately, it has not yet been thoroughly tested, and chemicals are still necessary for processing and laying. So the jury is still out on this material, but it is at least on the right track.

Several resource lists for nontoxic household items have been compiled and are circulating among environmental consultants and the general public. Both books by Debra Lynn Dadd are highly recommended in this regard, and are often found at health food stores.

Residential environmental specialists can often spot problems before they get out of hand. Before remodeling or moving to a new house, it is a good idea to hire one for a complete survey and report. This small investment at the beginning of a change can save countless dollars and huge amounts of misery later on. These specialists are still few and far between in America. The International Institute for Bau-Biologie and Ecology in Clearwater, Florida has certified over 30 Environmental Home Inspectors as of January 1995, and is set up to make good recommendations.

PART IV:
INNOVATIVE SOLUTIONS

CHANGING SUBTLE ENERGIES WITH FENG SHUI

INTRODUCTION

Specific definitions of Ch'i are sometimes difficult to understand as translated from Chinese documents. A translation saying, "That which cannot be defined is the essence of Ch'i" is not much help to modern researchers. That said, we can define Ch'i as a fundamental energy of the Universe, which creates electromagnetic and other energy patterns, and propagates much like sound.

Essentially, according to ancient documents and modern practice, Ch'i energy is regarded as the same thing in both acupuncture and Feng Shui. It can truly be said that "Feng Shui is to buildings what acupuncture is to human bodies." By manipulating the subtle energy patterns found in buildings, it may at times be possible to create healthier environments for people, and thus improve health. This is why Feng Shui is of concern to therapists.

RITUALISTIC FENG SHUI

Ritualistic techniques are those for which no working mechanism can be determined, such as purification rituals. Parts of traditional Feng Shui as practiced in China rely heavily on secret rituals. Some of this information has been made available through Master Lin Yun's Black Hat school in Berkeley, California. In some ways, these rituals are reminiscent of European ceremonial magic. Usually, the rituals need to be done only once, or for a specified period of time according to a schedule. It is possible that unknown mechanisms are at work. However, the secrecy surrounding the rituals makes a scientific evaluation impossible, until such time as a master is willing to allow everything to be revealed.

One way in which rituals could be valuable is in the purification of places. Several doctrines throughout the world say that certain events, especially violent deaths, can leave behind vibrations. The nature of Ch'i energy could easily explain these persistent vibrations. It is entirely possible that, through long experience, effective purification rituals have been developed which could change the Ch'i vibrations permanently through some mechanism embedded in the ritual. It has been noted that music, played live in a ceremonial context, sometimes appears to have a lasting effect.

MENTAL TECHNIQUES

Mental Feng Shui techniques are those which can be explained through an understanding of the mind. A good example is placement of mirrors near an entrance. This tends to raise self-esteem, since almost everyone gets a positive charge out of seeing themselves in a mirror.

In rural Kansas, a client was experiencing severe mental stress combined with a deteriorating marriage. During a survey of her house, several positive design elements were found. However, two decorations were definitely creating stress. One was a hayhook, suspended from the high point of the living room ceiling. Two sharp points were aimed down into the room. This kind of decorative motif will always cause stress. Using a farm implement for decoration seemed like a nice whimsical touch at the time, but its effects were even noticed by a ten-year old visitor to the house, who asked if they were afraid of the thing coming down on them. Truly, the points were constantly coming down on the occupants, causing a sense of uncertainty, even though a stout rope held the offending appliance securely in place.

The other problem was the placement of a gun rack next to the bedroom door. In rural Kansas, gun racks inside a house are normal, as predators sometimes need to be promptly dispatched. However, the barrels of the two guns were pointing straight across the threshold of the bedroom. This constantly transmitted a subliminal message, saying "marriage is dangerous" over and over. So the relationship deteriorated. As soon as these two items were moved, the clients reported a marked improvement both in stress levels and in the marriage.

Entrances to buildings are another example of a mental Feng Shui factor. If an entrance is blocked, the occupants tend to fall into thought patterns which block the flow of money or other good things into the house. Case histories involving improvements after clearing up entrances are too numerous to mention here. The front door has a great deal of meaning, even if another entrance is normally used. This is the face presented to the world. Family breakups or schizophrenia have often been reported where more than one front door is present, making the proper entrance unclear. These things will happen to every occupant of a house, until the situation is fixed.

Building shapes are another aspect of Feng Shui relating to mental processes. People tend to seek balance in life, and a building with an

128

unbalanced shape can indeed cause a corresponding unbalance in an area of consciousness. Each directional corner of a building has a subliminal association. These have been classified based on mathematical principles underlying the I Ching. Appropriate diagrams are available in almost every Feng Shui manual.

The legend on the water tower has said "Watch Nunn Grow" for over 30 years. Meanwhile, the population has remained steady. A vague, split main street works against this whole town, breaking up Ch'i flow to the extent that nothing ever happens here. Part of the main street is to the right, and part off to the left.

CH'I AFFECTING PRINCIPLES

Ch'i affecting techniques are designed to work directly on Ch'i energy, and are of the most interest to subtle energy researchers. Some of these techniques involve placing of crystals, which are suspected of having an effect on subtle energy patterns. Other techniques involve looking at water flow, to take full advantage of the associated Ch'i energy.

Much of Feng Shui is concerned with the proper speed of Ch'i flow. In a well-balanced environment, Ch'i can nurture all the space and the people in that space. It must move slowly enough to impart healthful energy to everyone there, and fast enough to go back out again into the world without stagnating.

In cases of fast-flowing Ch'i, known as "Sha Ch'i", it has been noted that mental and digestive disturbances are often associated with this phenomenon. The most obvious example of fast Ch'i is when automobiles constantly point towards a house in a cul-de-sac, a T intersection, or on a curved street. During childhood, the author lived in a house across from a T intersection for nine years, and a large number of the family's problems were later discovered to be typical of this arrangement. Paranoia, irritability, chronic anger, and heart dysfunction have all been associated with this configuration. Some interior design configurations, such as long hallways, can also create this. Fast Ch'i needs to be either diverted or broken up. Crystals, screens, plants, and ironwork can all be used as design elements to accomplish this, although in some cases there really is no good solution.

Stagnant Ch'i is associated with other kinds of problems, such as lasissitude, mental deterioration, respiratory dysfunction, and sometimes tumors. This can occur when a house is built in a former swamp. Bedrooms located too far from the rest of the house, where there is no chance for Ch'i to circulate properly, can also create stagnant conditions. Mirrors can be used to reflect Ch'i into stagnant areas. Sometimes, remodelling is called for, to integrate stranded rooms into the rest of a house, or place new windows to bring in Ch'i associated with light.

FENG SHUI AS AN ADJUNCT TO HEALING PRACTICE

Any time a patient visits a therapist for the first time, the therapist could order, as a standard procedure, a full Feng Shui evaluation of the patient's home. Recommendations by the Feng Shui consultant would be shared with the therapist, and the recommendations, which often include remodelling or redecorating, could be more consistently implemented because of follow-up suggestions by the therapist. This is a compelling vision, which brings together healing practice and everyday life. Therapists are often frustrated because an unknown environmental factor is subverting treatment, and Feng Shui professionals are often frustrated because recommendations are not implemented. When therapists and Feng Shui consultants work together, both problems can be solved at once.

Feng Shui consultants ought to be preparing full written reports from each site visit. Integrating Feng Shui with any healing practice makes this a necessity. Relying on memory or notes taken by clients is not reliable enough when integrating the recommendations into a healing regimen. Since each patient is essentially a living experiment, there should be a focus on methods to take case histories and create journals which will build up the body of anecdotal evidence in this discipline.

POSSIBLE METHODS OF MEASURING CH'I ENERGY

When working with Ch'i modifying techniques, a method of measurement would be a great advance. Right now, no consistent and reliable method of measuring Ch'i is known. Many subjective techniques are available, including several forms of kinesiology, radionic stickplates, and dowsing. All these methods depend heavily on the sensitivity and personal expertise of the individual making the evaluation. Since individual sensitivity to all subtle energies can vary from one day to the next, or even faster, it is best to continue seeking some form of consistent analog or digital measurement method.

Since Ch'i is said to propagate just like sound, this is an important possibility in developing measurement methods. Jack Derby of Tucson, Arizona, who built the Violet Ray Crystal Resonator device, found that one model of Radio Shack acoustic meter showed high readings when placed near people hooked up to the device. Since the device definitely seems to be affecting Ch'i energy, based on reports from many subjects who have tried it, this is an important line of inquiry.

It may be possible to couple radionic stickplate circuits to digital readouts. Coupling dowsing sensors to digital meters is another approach to try. After all, it doesn't matter what the numbers are measuring, as long as there is a consistent basis for comparison between sites and methods. We should be able to figure out the exact meaning of the numbers after sufficient data has been built up.

CONCLUSION

Feng Shui is only beginning to be understood in America and Europe. As understanding increases, we should be able to more consistently use its power to improve health and prosperity. Interfaces between health professionals and consultants need to be developed and strengthened. Traditional ritual information may not always be available, but many other aspects of this discipline are open to further study, correlation, and development. The most exciting area by far is the possibility of directly affecting subtle energies using Feng Shui techniques. As subtle energies are further classified over the next few years, we are likely to see tremendous advances in understanding old procedures and developing new ones.

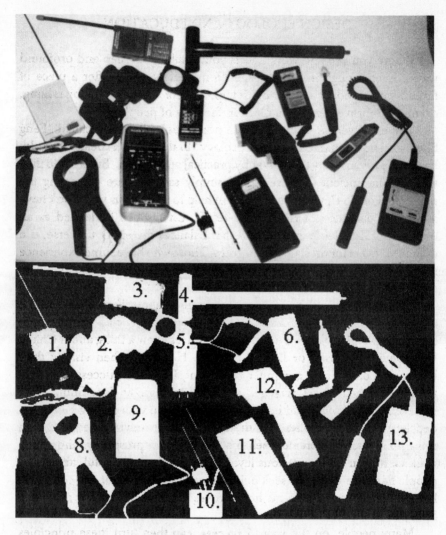

A DESIGN ECOLOGY TOOLKIT

1. Nichols All-frequency Monitor
2. Binoculars (for roof evaluation)
3. Walkie-Talkie radio
4. Merkl High-Frequency Antenna
5. Light Meter
6. Microwave Oven Seal Checker
7. Digital Thermometer
8. Ammeter Clamp Probe (for pipes)
9. Fluke 87 Digital Multimeter
10. Electric Field Antennas
11. Geiger Counter
12. Instant-read Digital Thermometer
13. Meda Gaussmeter

DESIGN ECOLOGY AND EDUCATION

Where you live and work affects your thinking in deep and profound ways. Even though it doesn't seem to make any sense, after a piece of furniture is moved, or a front door is improved, or a new decoration is hung, dramatic improvements take place in the lives of people.

The effectiveness of these design principles, collectively called "Feng Shui" and now renamed "Design Ecology" in the English-speaking world, has been proven many times over by practical application. But why do they work? The ancient Chinese would simply say, "we are correcting the movement of Ch'i", which is not a lot of help to those of us who don't have a working definition of this foreign term. Ch'i, which is defined as an invisible, undetectable energy flow that balances the entire universe, is a difficult concept for most people to grasp. Those who have some experience with mysticism find it easier to comprehend, but difficult to explain to anyone else.

Then, a few years ago, as more information about Feng Shui became available in the United States, I noticed some interesting things about successful people. None of these acquaintances knew a thing about China, Feng Shui, mysticism, or Design Ecology. However, when visiting their homes and offices as a computer consultant, I saw that successful people usually had their places arranged very well in terms of Feng Shui. Conversely, failing families and businesses usually lived with glaring Feng Shui deficiencies. This implies that there are, in fact, environmental design principles which can create a sense of balance and prosperity, and which work largely on a subconscious level. The fact that successful, actualized people have a tendency to seek out balanced places without any conscious knowledge of these design methods proves that Design Ecology and Feng Shui can work.

Many people, on the way to success, can then turn these principles around and manipulate design elements in order to get a boost toward their goals. Therefore, while educating yourself to a higher state of mind, it is possible to use your living and working environment to reinforce that programming. This is a powerful two-pronged approach. Either technique can work, as has been proven many times, but putting them together holds even more promise. It also prevents many of the more insidious forms of sabotage that students sometimes inflict on ourselves. We can make environmental design artifacts constantly serve as a reminder of the principles we're attempting to integrate.

133

FEELING IN CONTROL

One of my clients told me about a business he'd previously been involved with, which had folded. He said it had four partners, and each went their own way. From that information alone, I told him they had been in a common work area, with each partner using a desk in one corner of the room. Their backs were to each other. My client said, "how did you know that?".

A receptionist was having serious difficulty learning the new software at her job. Her boss liked her a lot, but was reluctantly considering letting her go. She sat at a desk facing a wall, with an L-shaped extension holding the computer equipment. When she turned to sit at the computer, her back was directly to the front door of the office. I suggested that she move her desk so that it faced directly to the door of the office. This caused a slight decrease in space in the reception area, but it rotated the computer extension so that it faced a side wall, instead of the back of the office. That way, when working at the computer, she could easily see people coming in. This was over three years ago, and she is still working at the same company, having won her battle with the new software shortly after her desk was moved.

Both of these examples illustrate principles often noted in Feng Shui manuals, and point out obvious mental effects of design. We all like to feel comfortable where we work and also where we sleep. If you extrapolate the above examples to the bedroom, it follows that a bed should have a good view of the door. This simply makes you feel more secure and in control.

THE EIGHT CORNERS

Sooner or later, when studying sources based on ancient Chinese science, you will run into a concept called the "Ba-Gua". There is no exact translation into English, but you could call it "Eight Corners" if you want. The foundation of this concept is that everyone has an instinctive desire to see things in a certain order.

According to Feng Shui consultant Thomas Howes of Napa, California, the human mind contains an imprinted map of the Ba-Gua characteristics. This could thus be called a neurolinguistic phenomenon. If a particular part of a building or room is blocked off, missing, or enhanced, results will consistently occur which manifest as behaviors in patterns corresponding to the observed physical characteristics. By explaining the phenomenon in neuro-linguistic terms, we have a way to deal with the concepts in an orderly manner which can be researched.

Accroding to Feng Shui theory, by enhancing a particular corner, you can increase the potential for that characteristic in your life. Conversely, when something decreases the feeling of a particular corner, it is the corresponding area of life that is then most susceptible to damage. Enhancements to a corner are positive pictures, harmonious colors, good views out windows, and sometimes mirrors (which tend to increase self-esteem). Detriments to a corner are negative or dark pictures, drab colors, views of unpleasant surroundings, and toilets (symbolic of things going down the drain). So Feng Shui is partly a process of finding appropriate enhancements and removing or covering up detriments according to this ancient map of the eight corners.

Each corner can correspond to a cardinal direction or, in more advanced schools, to a corner of a room or building, by considering the entrance as one of the three North directions. In other words, as soon as you enter a building or room, that's north right there, according to the standard orientation of the mind. The rest of the directions are extrapolated from that point. We will go through each of the eight corners, giving examples of ways that each corner can be enhanced for maximum effect, according to both traditional and modern experience. (Readers in the Southern Hemisphere will have to switch everything backwards to partially compensate. There are also other geobiological considerations operating there which further modify these characteristics.)

<u>North</u> We start here because this is the ideal entrance to a room or building, according to Feng Shui. Since it is associated with one's career, anything that adds excitement and vibrancy to this area of a room is good.

<u>Northeast</u> Knowledge is here. Many practitioners encourage clients to put books in this corner of a room, if possible, to create a harmonious balance. It can also be a good place for a computer. If the door is here, then use the door as a symbol of your own quest for knowledge.

<u>East</u> Family is what keeps the human race going. If it is damaged or lacking, inspiration is difficult to find for most people, and distorted thinking easily takes over. You may notice that many Asian people put pictures of ancestors in this area of a living room, to stress the continuity of a family through time.

<u>Southeast</u> In the winter, the sun rises here. This is symbolic of wealth, allowing us to physically continue with life. Some people like to hang firecrackers or representations of money in this corner of a room. It's a good area to use when reinforcing prosperity consciousness, by making it bright and cheerful.

<u>South</u> Fame or reputation. Near the San Francisco airport, there is a

hotel with a terrible reputation. It is ironic that you see a fountain along the "south" wall when entering the lobby, as if its fame were constantly being quenched. Since you could call this an active, even fiery quality, it stands to reason that water symbols are inappropriate here. Again, vibrant, alive decorations, or expansive views out a window, are best in this section.

Southwest This is marriage, or, more generally, a primary partnership. Here is where water symbols are good, invoking the ideas of flow and adjustment necessary for success in this area. Pictures of rivers and ocean scenes can be appropriate.

West Creativity and children are the characteristics of this corner. In ancient China, since children were so highly valued as a potential support for one's old age, this corner was often the most enhanced of the whole house. As you can imagine, complex factors are at play in this corner. We need to carefully consider what our goals are in regards to children or creativity, and carefully choose decorations that clearly suggest these goals.

Northwest Cooperation with others is the theme here, which can manifest as a business partnership, or as friendships. Ask yourself, "when I ask for help, do I get it?". If not, try enhancing this area with warm, positive pictures. Artwork or tapestries showing groups of people in happy, cooperative situations are especially good here.

PUTTING GOOD DESIGN INTO PRACTICE

You might call Design Ecology the art of healing the Earth, one building at a time. Of course, there's much more to good environmental design than just slapping pretty pictures and mirrors on walls. That's why some people are drawn to specialize in this field, spending years building up experience and doing extensive research to draw on the experience of others. However, by being observant, there is still much that each person can do to enhance an environment. In the process of considering the ideas put forth here, most people are going to accidentally find many other ways in which a home or workplace can be improved. This is an exciting process in itself.

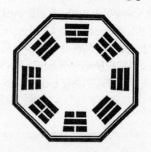

136

Geodesic Solar breakfast nook on Stuart Gordon's home in Silverthorne,
Colorado

ACTIVE SOLAR SYSTEMS

Active solar systems fall into several distinct categories. Not all of them
are particularly effective – many require too many materials and too much
maintenance to ever be economical. Others are excellent investments for
any homeowner, and some designs are especially good for businesses. First,
these are grouped by general function, and then by classifications within
each function.

FLUID HEATING

Hot water is needed for many household uses, and it can be an energy
drain on a house. Also, it is sometimes possible to design an effective solar
system using heat exchangers for space heating with fluids. These can be
relatively complicated, but they have worked in many situations. In general,
the main drawbacks to any fluid heating system are initial material costs and
maintenance needs.

Draindown or Drainback

This is the simplest kind of water heating system. It also tends to require the greatest amount of maintenance. The basic concept is that water is heated directly by a solar panel. Whenever temperatures in the solar panels are insufficient to provide heating, the water is simply drained out. Usually, specialized electronic controls have to be installed to make this kind of system work right. Also, because of temperature variances, piping in these systems needs periodic maintenance. If the controls ever fail, even once, in freezing weather, the whole system could be destroyed. Therefore, this kind of system is only recommended for tropical and semitropical climates, or as a supplemental pool heater in the summer.

Fluid Heat Exchange

To get around freezing temperatures, several types of fluid-charged systems were developed. That way, no drain controls are needed, and control issues can focus on heat provision. The most commonly used fluid is propylene glycol, a basically non-poisonous antifreeze sometimes used as a food preservative. Other types of antifreeze have been used as well. A silicone antifreeze was developed in the early 1980's, and was used in some brands of systems, but its advantages over propylene glycol were dubious relative to its expense.

These systems are usually used to heat domestic hot water for washing and bathing. They also can route either water or antifreeze into room heat exchangers. However, every additional run of piping to a heat exchanger reduces the efficiency of the system and adds another potential maintenance problem in the future.

Refrigerant-charged

These are really glorified heat pumps. They are mentioned because in the early 1980's several companies around the United States marketed these as solar systems by adding a solar collector to a heat pump. This can give the system a slight operating efficiency boost, but it is usually not worth the extra material costs. Electric power consumption on some of these is high. In spite of these drawbacks, there are still a few of these systems being marketed occasionally.

AIR HEATING

With active solar air heating, it is possible to create a simple but elegant system which produces a lot of heat relative to the investment. Field experience and calculations showed in many cases, an air system could be much more economical than a water system. Because space heating needs are greatest in the winter, it was found that air panels of all types could easily be mounted flush to a south-facing wall and still create a lot of benefits. This can significantly reduce material and labor costs by eliminating frames and brackets to hold panels at the proper angle.

Integral Wall Panels

An existing south wall of a building can be made into a large solar collector. These are sometimes referred to as Trombe Walls. A metal sheet, painted flat black, is tacked to buffer strips on the wall. Then, the sheet is enclosed with a frame and double glazing. If desired, extra insulation can be added behind the metal sheet. Two holes, for intake and output, are punched in the existing wall. The holes are usually made in standard duct sizes for convenience. An air moving fan, which can be thermostatically or manually controlled, is added to one side. Many of these exist in the San Luis Valley of Colorado, where low-income residents found them extremely practical and affordable in dealing with the many cold, sunny days there. Sometimes, several neighbors get together to build these on each others' houses.

Active Greenrooms

Sometimes, it's possible to build a small greenhouse addition to a house which can provide extra solar heat to the whole house. It is even possible to run a duct to another part of a house, where heat is most needed. Thermostatically controlled air fans make this an active air system. Be careful with these additions. Too much sunlight can overheat a house, or become oppressive in other ways.

Manufactured Panels

More efficiency can be gained from specially made air collectors. A large number of designs have been patented to make air gather the maximum amount of heat possible on its journey through the panel. The key to manufactured panels is actually reducing heat loss from the front. It's quite easy to insulate the back, but the front is another matter, since it must also be protected from hail, dust, and ice damage. Lexan and acrylic glazing were tried, which is good only in areas that aren't dusty, since they scratch easily. Several types of fiberglass, commonly used in greenhouses, may emit toxic gases. Glass can only be strengthened to a certain degree, but it may be the best choice in many instances. Double glazing is virtually mandatory. Any further layers of glazing reduce heat transmission unacceptibly.

PHOTOVOLTAIC – SOLAR ELECTRICITY

Many types of solar electric power panels are in use right now. The world's largest manufacturer is Siemens in Germany, and the second largest is currently Solarex. Several highly technical types of photovoltaic (PV) cells have been developed. At the moment, amorphous-silicon types are regarded as the most efficient overall, but that could literally change tomorrow as manufacturing processes change and research progresses.

Voltage standards are changing rapidly. Most PV systems were initially set up on 12 volts DC, to conform with recreational vehicles. Lately, several 24- and 48-volt DC systems have emerged. The higher voltages make inversion to 110 volts AC more efficient, so that computers and other normal appliances can be used for longer periods. Other technical advantages are available at the higher voltages as well.

Most systems consist of a series of PV panels, connecting wires, storage batteries, and a control center. Inverters and backup generators are optional. Some manufacturers, such as Ananda, are setting up integrated units which include inverters along with all control functions. Users can then simply plug in wires and start running.

When using PV power, it is best to carefully consider which appliances are necessary and which can be left out. High energy-using electric appliances such as space heaters, air conditioners, and dryers are out of the question. When possible, it is best to use as many appliances as possible which do not require inversion, which always wastes some power.

Except for the inverter, there are normally no parts of a DC electrical system which would cause potentially harmful electromagnetic fields. Inverters, when used, should always be located away from working or sleeping areas. Most models tested so far have put out strong fields extending several feet in every direction. Inverters are often necessary, as 12-volt appliances tend to be expensive and less durable than regular AC designs. Certain appliances aren't even available in DC versions at all. Computers generally need AC power to run, especially models with large monitors for desktop publishing and graphics work.

PV systems are an exciting area of new advances in technology. New innovations occur almost daily. Critical areas needing improvement are battery design, backup systems, and appliance design.

PERSONAL COMPUTER ENVIRONMENTAL CONSIDERATIONS

ELECTROMAGNETIC FIELDS AND OPERATORS

Many people are concerned about the effect of electromagnetic fields on anyone operating a computer. While sitting around in front of a computer screen is probably not as good for a person as working in a garden, the actual fields emanated by most modern equipment are not all that bad. This, of course, applies to measurable fields, and assumes that the equipment was made since 1988, when consumers began to demand low-emission equipment. Older machines may have high levels of emissions. As it turned out, designing for low emissions wasn't all that difficult, so most machines sold since 1988 are fairly safe.

There are a few things to watch out for when placing computers in a workspace. First, make sure that the back of a computer, where the electric cord comes in, is at least two feet from where anyone sits. Except in densely populated cubicle offices, this is usually not a problem. Many printers have high field levels coming from their control panels, so it is a good idea to have a printer at least two feet away from any part of the body. For many home computer systems, putting a printer on a shelf directly underneath the computer is a common practice. In this case, the control panel would be just a couple of inches from the operator's knee, which is not a good idea.

Watch out for power transformers at the plugs to some printers, modems, and other peripherals. Any box at the plug end of a power cord is transmitting a magnetic field for a short distance, and so should be at least one foot away from the body.

Most accessories which reduce electromagnetic fields really don't help a lot. There are some exceptions, such as the Ray-X Button by VrilTek, which uses innovative transformative materials. In fact, the Ray-X has proven to be the best device of its kind, especially relative to its low cost. Some especially sensitive people find they feel better wearing wrist straps which are connected to a ground. This is most effective if the ground is dedicated to the wrist strap, and does not interface with the building's electrical system at all. The majority of operators find wrist straps a nuisance.

Instances of health problems among computer operators usually have in common suppressive supervision practices and older terminal equipment. Any organization that tries to do things too cheaply and comes down hard on its workers can expect to have lots of problems, regardless of the electromagnetic environment.

EQUIPMENT LOCATION

Placing of computer equipment seems to be a common-sense procedure. You make sure the place is reasonably warm and dry, and set it all up. Most of the time this approach works well enough, especially for the short term. Sometimes, however, certain locations can cause persistent or intermittent equipment troubles that can bring an operation to its knees.

Electrical environments should always be considered. An electronic parts salesman was in the habit of carrying a notebook computer with him on sales calls. One day, he had to go to someone's back shop in order to look at their equipment. He took his computer along, to type in the specifications on site and save some time. On the way back there, he had to pass between two giant 300 Kv power transformers. When he turned the computer on, he found that every bit of data on his disks was gone, along with the computer's ROM chip programming. Direct damage amounted to over $1,400, not counting the lost disk data.

Most environments are not as extreme as that, but it is an instructive example. Even smaller, transient atmospheric fields can occasionally cause damage, and so a good general rule is to locate computers as far away as possible from power transformers and similar heavy electrical fields. From an operator standpoint, this is also a good idea, since there is existing research data indicating that exposure to heavy electrical fields over a period of time is not healthy. Sometimes, power transformers can be hidden in walls in commercial buildings, so having a survey done is appropriate.

Contrary to the old cliche, lightning can strike twice in the same place. That's because, since it always "looks" for the easiest path to ground, it will often go along the same path many times in a particular area. Even being near a lightning strike could be traumatic for a computer. Lightning is, therefore, something to consider when locating a computer. If you are in an area where lightning strikes have been observed to occur often, then heavy-duty surge protection and an Uninterruptible Power Supply (UPS) need to be installed. Lightning rods can also be a good investment in some areas.

It is even possible for lightning to destroy equipment if it only strikes near, and not on, a building. In July, 1992, an air burst over the house I was living in blew out an oven, the garage door opener, a tape recorder motor, and several other appliances. The computers had been unplugged at the first sign of a thunderstorm, but a telephone, modem, and a communications card inside the PC were blown through the phone line wire. There was no evidence whatsoever that lightning had actually struck the house. All of these effects were thus purely inductive.

In another case, a consultant lost a computer which had been unplugged during a thunderstorm. A length of printer cable was sitting underneath the computer. Since it was in a coil, an instantaneous field created by the lightning wiped out the computer, and melted the cable!

Temperature fluctuations can sometimes be a problem. There have been cases of computer chips becoming unstable below 40 degrees F. In a workspace that is unheated at night, all kinds of strange, intermittent behavior can happen for the first hour after a machine is turned on in the morning. One solution is to leave the machine(s) on all night, which is not as bad as it sounds, since PCs don't really use very much power. They typically use less than 1/10th of the power that a small space heater might use. Arctic environments, of course, demand special consideration and usually special equipment built to military specifications.

DEALING WITH BAD DAYS

Everyone has had "one of those days" at some time or another. While this may seem to be part of ordinary life, observations have shown that, in many cases, a high proportion of people in a particular locality have "one of those days" on the same day. When all of this begins to affect people who have microcomputers, very strange disasters can result, and, all too often, the reaction of people to these disasters causes more problems.

Here is where the controversial "ion field theories" come into consideration. Most storms (even dry ones) will be preceded by what are called positive ion fields. This is electrical shorthand for a lack of free electrons in the environment, which can affect both people and microchips. In microchips, a lack of free electrons can cause random current fluctuations that manifest as unaccountable data errors. Peripherals such as switch boxes that often are built with cheap chips are especially prone to this kind of problem. People under the influence of a positive ion field have been observed to become more irritable, less able to think clearly, and more prone to random operator errors.

In other words, it is perfectly normal for some pieces of equipment to manifest random failures before a storm, and it is perfectly normal for people to compound the problem by panicking during the early stages. Typically, positive ion field patterns occur two to three days in advance of a storm, and again two to four hours before local precipitation occurs. Sometimes, even more dramatic ion fluctuations occur during drought periods, when so-called "dry storms" pass through an area. In the West, these are sometimes called "thundersprinkles". One of the most frustrating things about this kind of storm is that, unlike a normal rain or snow storm that clears the air of destructive fields, these often leave behind another positive ion cloud, which can cause equipment failures and erratic personnel behavior to persist.

There is only one effective way to deal with ion fields. Managers and lead operators must train themselves to remain calm and retain a sense of humor whenever any equipment failure or operator error becomes apparent. Then, if further problems are experienced, leaders must delicately balance humorous statements with direction that handles the current crisis, while still allowing other work to proceed. The best managers and lead operators, then, are either people who are highly sensitive to ion field fluctuations and can recognize instantly the signs of field-induced crises, or people who are totally insensitive to these fluctuations and can maintain a constant positive attitude no matter what is going on around them.

LOCAL AREA NETWORKS

When you string together a bunch of computers, modems, and printers with various kinds of cable and odd boxes, the potential problems multiply. All environmental considerations for individual PC installations must be met, and then all cabling must be routed through proper environments and shielded if necessary.

One early LAN installation was plagued with intermittent data interruptions, several times a day, causing serious problems for all the operators there. Repeated re-installation of the network software didn't cure the problem. Finally, one day a technician, having run out of other things to check, traced all the cable runs. The unshielded twisted-pair wire at one point ran behind a pop machine. Right then, someone walked into the break room and bought a soda. Immediately after that, an operator was calling for the technician. It turned out that every time somebody bought a soda, the slug-detector electromagnet put out a strong enough field to disrupt anything being transmitted on the nearby cable.

This is why shielded cable is a must for any environment where intermittent electromagnetic fields are a possibility. Even the ballast in an ordinary flourescent fixture could be a potential source of disruption.

Many switch boxes and other peripherals are built with cheap chips. That can cause other problems. An ion or electrical disturbance that doesn't affect well-built servers and workstations can sometimes play havoc with lower-quality devices. Any time a network is experiencing intermittent switch-related failures, switch boxes are a good place to start looking for causes.

Since most networks are complex beasts, environmental disturbances that cause data problems are very difficult to evaluate. Network administrators need to be aware that weather patterns and magnetic disturbances can be a factor, and should learn to look to those possibilities when no other cause is immediately apparent. Of course, good personnel management practices, especially during crises, are a must.

BAU-BIOLOGIE AND WEATHER

Three kinds of invisible fields typically ride along with weather events such as storms. These have been classified as: Electromagnetic, Ion, and Ch'i. Apparently, Ch'i fields create the other two types in some manner, which seems similar to other cosmic interactions postulated in physics based on the work of Nikola Tesla. For a more detailed explanation of these three fields, please refer to the chapter on "Revised Meteorological Nomenclature".

According to BauBiologie, effects of the three invisible weather fields can be magnified due to building location and design. In this short summary, we will consider the main points of this interaction.

BUILDING LOCATION

Any building located over disturbed areas will tend to catch more invisible weather fields than normal. We know that lightning often strikes repeatedly in certain places. If a building has unfortunately been built over one of these places, there will always be problems with electrical and computer systems because of lightning strikes. Also, static electricity buildup will tend to be higher in these locations. Therefore, frequent lightning discharge in and around a building becomes a useful indicator of a disturbed zone, and can allow us to make use of standard recommendations relative to geobiological (especially geomagnetic) disturbances.

In addition, carpets in buildings located over disturbed areas need special attention. Most modern carpet fibers cause intense buildup of static electricity, which can bother many people. Therefore, if a geomagnetic disturbance is found, it becomes especially important to make sure only natural fiber carpets are used in the area.

According to classical Chinese Feng Shui, Ch'i disturbances can cause subtle and longlasting problems around a building. From the context of ancient original documents, there are several different types of Ch'i forces, some of which have nothing to do with magnetic fields. These kinds of

145

disturbances normally cannot be found with instrumentation, so we must rely on dowsing to find them. Ch'i fields riding with storms can sometimes affect dowsing indications in a particular location as well. Thats why, when presented with a difficult location, its best to make a point of dowsing at several times, and include a period as a storm approaches in the schedule. This can provide valuable extra information as the storm Ch'i field interacts with the field on site.

ION FIELDS AND INTERNAL BUILDING SYSTEMS

Certain types of heating and air conditioning systems tend to create intense positive ion fields or electron depletion regions. Specifically, the worst culprits tend to be electric radiant heaters. Gas forced air systems tend to be fairly bad as well. As storms approach, the related ion fields can create more intense reactions in buildings outfitted with these systems. In regions like Colorado, Southern California, Israel, and Switzerland, where electron depletion events are often connected with several different weather event types, it is best to completely avoid using these kinds of heating systems. In commercial buildings, the likelihood of computer system failures connected with ion events makes it financially worthwhile to replace these heating systems.

Since weather patterns can sometimes be a diagnostic indicator of geomagnetic or geobiological disturbances, internal building systems can be reconfigured based on these observations. Electrical and computer systems can then have extra protection added.

CONCLUSION

Even though weather fields are transient in nature, persistent phenomena associated with them can be observed at specific sites and applied to BauBiologie. Further experience with this field will certainly produce more useful correlations between invisible weather fields and building systems. As instrumentation is developed to measure C'hi fields, this will become even more important.

MEDICINE WHEELS IN YOUR BACK YARD

Over 40,000 Medicine Wheels were found in North America after the White People started migrating in. Most of these were destroyed, after being called "relics of the Devil" by preachers. Some in remote locations or on private lands were preserved.

It has been said a Medicine Wheel will balance energy around it. Apparently, most of the old ones were located on or near power spots, and were used as guides for ceremonies, vision quests, and travel. According to one old Cheyenne chief, "Medicine Wheels were put here before our ancestors came, to teach us how to build our tipis."

There are many different designs for a Medicine Wheel. They range from very simple, indicating only the four directions, to extremely complex, with 28 spokes and seasonal indicators for important stars such as Sirius or the Pleiades.

Now that White People have mostly overrun this continent, almost everyone is ideally supposed to live in a fixed location. This seems to be against the natural pattern of the continent, which was sometimes called "Turtle Island" because it is always moving slowly (currently westward at about three inches per year). It was said that local energy fields were always moving as well, making a place that would be healthy to live in one year troublesome the next. People living on this continent thus usually created mobile societies, which were inherently contrary to European concepts. Those who are now living in settled patterns have to make the best of the situation. People who are aware of the balance of Nature, and who wish to do something in their own lives to enhance that balance, sometimes have a desire to build a Medicine Wheel.

How one of these artifacts actually works is open to speculation. It may be that Medicine Wheels somehow channel Ch'i energy, which might be identified with tachyon beams or etheric lines of force in some Radionic systems. Gardeners have sometimes noted that having a Medicine Wheel nearby seems to help increase plant growth and health. Some people feel that a Wheel helps them retain a balance with the Earth, and thus makes them more effective in accomplishing their chosen life's work. Standing in one can be a powerful experience for a psychically active person, and has no discernible effect at all on others.

You will notice that all the effects mentioned here are non-specific. That's because any work with Ch'i-type energies will have different effects for different individuals, depending on genetic makeup, health situations, personal attitudes, and education. So to say that building a Medicine Wheel would definitely cure some problem would be wrong. But if you build one, and a problem happens to clear up, then that's fine.

Locating a Medicine Wheel is somewhat similar to locating a business. Some places should not have one. In general, current or former swamps are not good. The very highest place in an area is not good either. Deep in thickly wooded places are also poor locations for Wheels. Many existing Wheels are located somewhat off a high point, with the point clearly visible from the center of the Wheel.

It is conceivable that building a Medicine Wheel in the wrong spot could cause damage. That's why New Agers and others who are not absolutely clear about their purpose in building one, and who may carelessly locate it, should be discouraged from trying until they have learned more.

Medicine Wheels should be on as level a spot as possible. Dowsing for the right spot can be a good technique, if you already have that skill. Many urban or suburban back yards, if they are not too thickly wooded, and have a good level area, are good choices. It is important, however, to make sure that other energy considerations are all right. When in doubt as to building a Medicine Wheel in a heavily settled area, consult a Feng Shui practitioner, a Native American medicine person, or an advanced geobiology consultant. Rural spots are better, partly because there is a wider choice of locations. Ideally, a Medicine Wheel should be in a place with a good view, where energy can flow downward from it in at least one direction. Of course, sincerely asking the Great Spirit for permission and guidance in this matter is an excellent thing to do.

The simplest Wheels contain one stone for each direction, two between each directional point, one in the center, and two stones in each of four spokes toward the center. The center stone should be a special one special to the builder, either having come from some sacred location (with permission) or having been used in an important ritual.

More complex designs have a total of 28 stones arrayed around the outside, with rays going inward at 8 places, and a circle in the middle with several stones in a sort of cairn. Stonehenge, one of the World's largest Medicine Wheels, is also a good general plan for a large Wheel, substituting stones for each archway from the Stonehenge plan. Here is yet another similarity between Celtic and Native American practices — they often seem to fit together well.

Sometimes a temporary Wheel will do, using small rocks or even gravel pieces. You may not get the overall balancing effects that a permanent Wheel would bring, but it might be a help during a period of crisis. In one case, a Medicine Wheel was built out of pennies on the back porch of a troubled house! It did seem to help tone down some of the heavy emotional energies flowing in the place, although no dramatic effects were observed.

Medicine Wheels are a personal matter which can affect the surrounding. Any time one is built, it must be done with an attitude of respect and reverence. A lot of things in life are like that, so Medicine Wheels are a useful practice from that standpoint.

LIGHTING

Sometimes, changes in lighting have a profound effect on health. This can be due to electromagnetic fields hitting sensitive people, or because the quality of light itself can affect health. Many researchers say natural daylight is the best thing for people, and therefore artificial lighting should approximate that condition as much as possible for optimum health. There are three kinds of light fixtures in common use, each with its own advantages and disadvantages.

INCANDESCENT

These are basic light bulbs, which are the most common lighting in homes. Their output tends to be weighted towards the lower, red and yellow end of the spectrum. Generally, their light seems to cause few obvious health problems in most people, although there are exceptions. Incandescent lights are not good for plants. Some people prefer higher levels of light which are not generally available from incandescent sources. In high northern latitudes, however, it is more important to approximate normal daylight as much as possible. People suffering from seasonal depression definitely need something stronger than ordinary incandescent lighting.

Many incandescent lamps put out high electric fields, so it is a good precaution to make sure you're at least two or three feet away from the lamp. Actually, only measurement can tell if this is a problem, because several factors can contribute to the field's extent and strength.

The biggest disadvantage of this type of lighting is that it consumes a lot of energy relative to the amount of light produced. Calculations indicate that if every house in the United States switched to more efficient lighting, which usually means fluorescent, the energy savings would exceed the amount produced by all of our nuclear plants.

FLUORESCENT

Most businesses use this type of lighting, because it consumes a relatively small amount of energy while providing high levels of light. All fluorescent fixtures consist of two parts, both of which affect light quality and the electromagnetic environment. Tubes (or, in some newer applications, bulbs) contain a phosphorescent coating that glows and provides the light.

Ballast is usually hidden away behind the tubes somewhere, and controls electrical current to provide precise characteristics to make the tube glow.

Disadvantages center around the facts that not all coatings are equal, and that ballasts often emit large amounts of electromagnetic radiation. In many cases, the cost, time, and trouble associated with maintenance can wipe out any energy savings created by using fluorescent lighting.

Tube coatings are the secret to the different characteristics of fluorescent lights. Some tubes are designed to emit an approximation of daylight, while others are skewed towards one color or another for special purposes. By far the most common type is "Cool White", which evolved in the 1950's. While it provides an adequate amount of light, several types developed more recently, which are closer to normal daylight, are considered by many researchers to be healthier. When shopping, ask for a color temperature of around 5700K. While this is technically not full daylight, it is pleasant and healthful.

When tube coatings wear out, which is usually after about a year in business applications, the quality of light takes a definite turn for the worse. Any time a tube is seen to flicker or has dark spots on either end, replace it. Light coming from that tube is probably not healthful. Always replace all tubes in a fixture at the same time, even if some seem to be ok. This is due to the fact that if one has gone bad, the others are sure to follow shortly. Also, there can be hidden damage in tubes that look good to inexperienced observers.

There are cases where ballast can emit such a strong electromagnetic field that health effects manifest in sensitive people. Whenever you hear noise or feel excessive heat coming from ballast, it should be replaced as soon as possible, because you can be sure that its electromagnetic fields are out of control, affecting both the invisible environment and light quality. One quick way to find a bad ballast is to use an infrared thermometer, of the type used for energy audits, and point it at each fixture in a bank. If one of the fixtures has a significantly higher temperature reading than the others, that's a bad one. When ballast wears out, it will burn bulbs out quickly. Ballast can usually be expected to last about four years.

Some cheaper ballast types emit unsafe field levels even when new. If electromagnetic fields from the fixtures measure over 2.0 milligauss at head level, ballast is probably the culprit. When buying ballast, see if you can get the store to show you some that's hooked up, and check the field level with your personal meter, if you have one.

Because of electromagnetic field problems, it is best to keep fluorescent fixtures as far away from people as possible. In office buildings, people can sometimes be affected by lights from the floor below.

Recently, several types of electronic ballasts have been developed. A few of these emit dangerous levels of high-frequency radio waves, but most are good. These have to be checked on a case-by-case basis. Another interesting innovation is the use of DC wiring for an entire light circuit. This means only one rectifier has to be used, which can be deliberately located safely away from people.

MERCURY OR SODIUM VAPOR

Most street lights are of this type. They are usually not good for health, because the quality of light is incomplete and inconsistent. Some are quite noisy. Many can even work after they've been partially broken, but their light and noise emissions are worse than before.

Always block off your bedroom windows from street lights. Having any kind of bright light shining on you while sleeping can disrupt metabolism. When you consider that most street lights emit an incomplete spectrum, the possible effects are not good. Actually, street lights are most effective for showing criminals what's worth stealing.

ODD SCHEDULES

People who work night shifts often experience constant metabolic disruption. Studies have shown that night workers do best if given work lighting close to natural daylight, and then complete darkness in sleeping quarters during the day.

COMMENTS ON SMALL FLUORESCENT BULB TESTS

Many people wonder if the new capsule flourescent bulbs are a good choice. Thanks to people at Jade Mountain, an alternative-energy supplier in Boulder, Colorado, measurements could be easily made. They have a unique display board set up with most of the prominent makes of small fluorescent bulbs available for evaluation. These bulbs generally have bulb and ballast in a single unit, although some are separable. All of them use much less energy than conventional incandescent bulbs, and somewhat less than normal fluorescent fixtures. Also tested was one brand of electronic ballast, which has a longer life cycle than conventional ballasts, and uses less energy.

For the tests, meter probes were held about 12 inches away from each light. In normal usage, few people will ever get that close to a light.

Test conditions were not ideal, mainly because of the presence of an electric power meter on the display board. This is handy for seeing how much electricity is being drawn by each bulb or combination of units. However, it does generate its own magnetic field, up to about 6.5 mG at nine inches away, and it couldn't be shunted out of the circuit in the time allotted. To compensate for this, the probes were sometimes oriented at a less than ideal angle. The PL Outdoor, Osram Dulux 15 watt, and the Motorola Ballast were most directly affected because of their positions on the board layout, and, as the measurements attest, results were still favorable.

SMALL FLUORESCENT BULB TESTS

MANUFACTURER	POWER (Watts)	GAUSS-F	GAUSS-UF	V/m Back	V/m Meas	V/m JUMP
BACKGROUND--->		0	0.067			
Panasonic	27	0.1	0	5.4	11.1	5.7
Lights of America	27	0.1	0	5.1	11.7	6.6
Osram Dulux EL	7	0	0	6.0	11.1	5.1
	11	0.1	0	6.4	9.2	2.8
	15	0	0	12.0	13.1	1.1
	20	0	0	9.3	11.1	1.8
Orion PL	9	0.8	0.172	3.4	13.5	10.1
Osram ELR	11	0	0	2.2	4.6	2.4
	15	0	0	1.8	6.4	4.6
Panasonic Capsules	14	0.1	0.216	5.0	6.8	1.8
Philips Earthlight	18	0	0	3.7	9.6	5.9
PL Outdoor	7	0.5	0.394	6.5	11.0	4.5
Motorola: Electronic Ballast		0	0	5.4	13.4	8.0

NOTES:
    Orion PL spikes to .738 when turned off
    Panasonic Capsules spikes to 1.2 when turned on & off
    PL Outdoor spikes somewhat when turned on & off

    TESTING DONE 1-12-93 by Michael Riversong, QI Consulting

TEST EQUIPMENT:

Three meters were used in this test:

1. GAUSS-F: 60Hz filtered Gaussmeter by Integrity Electronics

2. GAUSS-UF: Unfiltered custom Gaussmeter by Super Science. Where zeros are shown in the data, there was no variation from background levels with the bulbs turned on.

3. V/m: Electro-Stress Meter by International Institute for Bau-Biologie and Ecology, with v/meter antenna attached. The most important measurement here is the V/m JUMP, which shows how much the air is saturated with electric potential around the bulbs. The variations in V/m Back were background conditions of the test, probably attributable to the power meter on the board. V/m Meas is the reading after each bulb was turned on. This

measurement is much more difficult than straight magnetic field measurements, and is subject to variations because of many factors. Measurements are in Volts per Meter.

CONCLUSIONS

These bulbs (and the electronic ballast) are very good from the electromagnetic standpoint. They all generate much smaller electromagnetic fields than conventional fluorescent fixtures. Thus, they could be recommended for use by people who are electromagnetically sensitive. The V/m Jump showed the most variation, and so can be considered when selecting bulbs. V/m measurements fell off quickly as the meter was moved away from the board in all cases, so at normal reading distances these bulbs can be considered safe for that parameter.

EXPERIMENTAL FIELD MODULATORS

Numerous devices have appeared in the marketplace which purport to alter environmental energies for better health and well-being. They have varying degrees of complexity, ranging from simple wire shapes enclosed in cases to sophisticated multifunctional electronic circuits. Some devices rely on crystals to produce an effect. Others use shapes derived from ancient mystical sources. Some devices may be outright frauds. Others may have effects which are not measurable through any known techniques. It is possible that a device could work well in one location but not in another. Sometimes, the fact that a device seems to work may just be a psychosomatic phenomenon. This short survey only covers a few devices, to give you an idea of what's currently happening in this field.

POWER CONES

Variations on this design have been in existence for many years. Typically, they cost around US$300, and appear to be hollow plastic cones about three inches high. They are supposed to be placed around a house to neutralize the effects of electromagnetic fields, cosmic rays, and geobiological phenomena.

When opened up, the ones I have seen simply had about 10 inches of bare, thin copper wire wound in a spiral around the inside of the cone. That's all. While it is possible that such a wire shape might have an effect, the price does not seem to be justified.

METAFORMS

Greg and Gail Hoag have created some attractive sculptures which can have positive effects on any building environment. They've used Synergetic principles in developing their designs, which are mostly intended to be suspended either over beds or in good decorative points. All of their forms are appealing to the eye, but because of their geometric design, including the judicious use of color in some models, they have potential effects on subtle energy levels. I often send their catalog to clients. If one of their designs appeals to you, it will probably work well, providing an extra boost to any space. Any environment where I've seen one placed felt excellent. If you don't feel comfortable with subtle energy concepts, the artistic value of these modules is unimpeachable.

Greg Hoag with one
of his Metaforms

CLARUS & COHERENCE UNITS

A number of reports as to the effectiveness of these units have circulated in California, Colorado, Iowa, and Florida. Usually, they are electric digital clocks with extra components and programming added to the circuits. They sell for between US$200 and US$500, depending on the model. Some are said to have a "personality", which makes them the more expensive models. These units with personality are recommended by the manufacturer for use in spiritual development, including meditation practices.

Extensive testing has been carried out on a few sample units. In April, 1993, tests were done to see if there were effects on magnetic fields, space charge, oscilloscope waveforms, phase angles, and harmonics. What we found was that no measurable electromagnetic parameter was affected when they were plugged into an electrical circuit.

This does not mean that these devices are ineffective. It simply means that whatever effect they have is not measurable with currently available technology. We can't discount the number of positive reports submitted by users. Many people have said they feel better, they've stopped having headaches, or their home life was more harmonious once the devices were in place. Sometimes, it took a few days for the effects to take hold.

I have personally witnessed one of these units (a Coherence ES-1 in 1991) absolutely remove an obnoxious radio signal in a residential electrical circuit that was interfering with my recording equipment. So there is something to this technology, but we don't know what it is yet.

SE-5

This is the most well-known Radionics machine currently available. It usually comes with a small pocket computer. The combination of the radionic circuit and the computer is said to generate subtle (and of course unmeasurable) energies which potentially could, at certain rates, clear negative energies from rooms and even whole buildings. A cheaper unit known as the SA-2 is available, which can broadcast a single rate in a personal biofield. It uses the same essential core circuitry.

These are said to be effective in many cases. However, programming takes a lot of time and experience. Also, there are cases where they have not worked at all. I often think of a therapist I know who tried to use one to improve his love life. Instead of improving, he ended up with an extremely unhealthy obsession with a certain woman. This obsession was threatening his whole career at the time I got disgusted and broke off contact with him. So the SE-5 is definitely not infallible, although it is good for carefully conducted experiments in this field. It is available through a number of sources.

SPACECRAFTER

Peter Lindemann's invention is a breakthrough for environmental consultants. Many years of careful research into Orgone energy, weather patterns, quantum physics, and harmonics went into the design of this device. It sends two kinds of subtle energy vibrations through a room, in an orderly progressive sequence. The controls are simple, and there is one variable control for field strength. You can select either "Pulse" or "Pattern" mode. In Pulse mode, the Spacecrafter is designed to clear residues of electromagnetic field patterns, sending signals out through a hemispherical quartz antenna. In Pattern mode, higher-level subtle energies are cleared. You can run a long or short cycle. The long cycle is one and 1/2 hours, and the short cycle is 45 minutes. A flashing light gives an indication of the stage of running. Flashes start off fast, and slow down in stages as the cycle progresses.

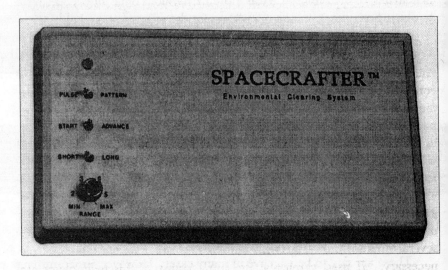

In my own tests, interesting phenomena have occurred around the device. The first place I used it was a house with five people and several dogs living there. The landlord was a relatively unstable person who absolutely refused to do any maintenance work on the place, including fixing the refrigerator. Ordinarily, this would have been a tense situation, especially given the personalities involved. Lots of food was spoiling, we were being threatened with eviction or a lockout, and all of us were from divergent backgrounds with widely differing ways of dealing with the world. But everyone treated the situation with unusual good humor. Four of the five people involved found new places immediately, in a virtually miraculous manner, and the fifth was willing to continue living there, fighting the landlord. The Spacecrafter was used almost daily throughout this period. I was the only person in the house who knew anything about Radionics or subtle energies, I said very little about the device, and so a psychosomatic effect was not likely.

Experiments in buildings said to be haunted by ghosts have been effective so far. That kind of energy cannot stand up against the field generated by the Spacecrafter. Any field not welcome in a building will be forced to leave quickly.

It is recommended that people leave a building when the Spacecrafter is running. I have seen no ill effects the few times it was necessary for me to stay in a building during a run, but it is better to be cautious until we understand these fields more fully. Cats have certainly noticed the device when it is running, and while their actions have been impossible to decode, they do react to it.

Once, the Spacecrafter was used to break down a tornado condition in a small Kansas town. Before tornados happen, the air gets extremely still, animals act nervous, and clouds have a distinctive color and flatness. As the condition was deepening, my friend and I decided to try it. I set the unit on the front porch (which faced west), cranked the Range knob all the way up, put it on a short Pattern cycle, and fired it off. Within ten minutes, the stillness in the air had broken up, and a pleasant breeze began to blow. Lightning was crashing on every side of the town, but not in the immediate area. Rain fell intermittently through the evening. The most startling thing was, when we turned on the TV and watched the radar track of the storm, there was a neat circle around the town, where the storm suddenly got lighter as it came our way. This is not something to try casually. My friend had been through several tornados and felt this was a time when weather modification was truly necessary. If used carelessly, the implications of this technology are frightening. Within a building, walls appear to limit the extent of the effect.

Occasionally, I have used the Spacecrafter in conjunction with an environmental survey. The clients involved have consistently reported that the space felt much better after the unit was run. It has proven useful in many situations. The effect is subtle, and probably unmeasurable, but enough people have reported positive changes in essentially blind tests where I feel confident in the Spacecrafter as an effective environmental field modulator.

HOUSE PLANTS

Simple house plants can effectively modulate an environment. It is now known that several species are capable of actually removing chemical, aerosol, and even fungal pollutants from room air. Many of the popular plants used in the United States have proven effective in this regard. They can be especially good in environments where pollutants are barely detectable but irritating. Obviously, there are limits to the amount of abuse they can take, but overall this is a good solution for many homes.

Recently, a practical guidebook on this subject has become available. It is called <u>A Breath of Fresh Air</u>, by LaRonna DeBraak. Specific species are covered, with useful notes as to their effectiveness and how to care for them. The publisher's address is given in the Resource section.

PRACTICAL APPLICATIONS IN SYNERGETICS

Buckminster Fuller's reconstruction of geometry, which he called Synergetics, has been absolutely incomprehensible to many. However, it has serious practical applications which can be uncovered by diligent attention to his work.

Nobody should ever attempt to read either of the Synergetics books straight through from cover to cover. These books, in themselves, are a synergetic structure and in fact it is best to have both volumes, rather than just the one that's easy enough to find. <u>Synergetics Two</u> is almost impossible to find, but the search will be well worth it for the serious student, because the second volume explains many problems and difficult spots in the first book. In addition, the newly released <u>Cosmography</u> book goes a long way toward simplifying some of the concepts presented in the Synergetics material.

The best way to read these books is to thoroughly study the chapter headings and section guides in the fronts of each book. With those in mind, one can pursue lines of thought weaving throughout the book in a perfect geodesic pattern. This makes the study intensely personal. Since Buckminster Fuller made up a lot of his voabulary, you will constantly run into unfamiliar words. He had hoped that these new words would be self-defining. This means that you have to stop reading when encountering these words and carefully consider what the unknown word could mean. It is hoped that a specialized Synergetics dictionary will eventually become available, to assist students.

The first thing these books do is to completely destroy any pre-existing notions concerning plane geometry and calculus. These old forms are built upon the assumption that things can be conceived of in terms of flat planes and straight lines, when, in fact, neither exist anywhere in the Universe in any observed experience. Fuller repeatedly stated that we had been using these tools for too long and need to progress beyond them as rapidly as possible. Creating a firm mathematical foundation which can be applied to any other science is the core function of Synergetics. This was always the function of mathematics in the past. For example, the discoveries of Isaac Newton and Galileo were fundamental to the evolution of civil engineering practice in Europe and America. Now, the discoveries of Synergetics can become fundamental to the evolution of ephemeralization, which is the process of "constantly learning to do more with fewer materials."

STRUCTURAL DESIGN

The most obvious use of Synergetics to date has been in the process of designing more efficient structures. The geodesic dome is the most well-known example. It is through Synergetics that the structural design of these domes was developed, because angles of the supports became easy to calculate. Geodesic domes use a minimum amount of materials to enclose a maximum amount of space. Thus they have become widely used in military applications. Applications for shelter have been a little more difficult to develop due to the fact that the multi-faceted surface of geodesic domes is difficult to seal from the elements. As that problem is surmounted, we find residential geodesic domes becoming much more common.

Another type of shelter inspired by Synergetics is now being promoted by Will and Cathy Sawyer in Maui, Hawaii. This shelter makes use of a single star coupling at the top of the structure from which radiate five supports. These five supports serve as a place to drape any kind of covering, be it plywood, plastic, canvas or salvage material. Simple triangular supports terminating at the upper radiating beams complete the structure and form stable walls. According to the Sawyers, this type of structure uses one-fifth of the materials that a conventional box house uses to enclose the same amount of space. One can be erected in less than a day. A similar concept has been seen in photographs of campgrounds in Central Europe.

Existing structures can be surveyed and, sometimes, optimized using Synergetic principles. The ideal angles postulated in Synergetics can be applied to an understanding of the harmonics of any structure. Once one gets used to thinking in terms of "in and out" rather than "up and down", we find that structures can even be used to enhance the intelligence of inhabitants. By understanding the mathematical process, remodeling recommendations can be made simpler and less expensive. For example, it can become possible to observe stress lines in a building and, without benefit of complicated engineering equations, instantly comprehend the optimum path for supplemental support. Through Synergetics, this can become an automatic mental process, allowing engineers of the future to spend more time on design optimization rather than simple structural calculations.

"Earthship" houses have attracted a lot of attention, because they use recycled materials such as tires, blend in well with the landscape, and can be built cheaply. Unfortunately, they are so labor-intensive, that many are started, but few get finished.

GEOBIOLOGY

People who live in conflict with geobiological patterns are more likely to suffer from certain illnesses, including cancer. This is according to

160

statistical research done in Germany, Austria and France over the past 100 years. Synergetics may be the key to unlocking the secrets of Geobiology. Therefore, we may find a prevention for cancer within the mathematical process of Synergetics.

At a certain wavelength of light, Russian satellites reportedly discovered a subtle pattern of lines extending all across the globe in a pattern very much like many of the illustrations in the later parts of the Synergetics books. These are some sort of structural manifestation, but we can't determine the exact nature of the structure at this point. It may be composed of energy waves corresponding to what the Chinese would call Ch'i power. According to the Chinese, this mysterious Ch'i power holds the entire Universe together.

These large grid lines have been seen criss-crossing the entire globe, and often converge near places of sacred significance according to native traditions. The practical application of these lines has been a bit unclear. However, in ancient geobiological practice, sometimes known as Geomancy, there were ways to detect what are known as "ley lines" running throughout the countryside. People could locate alongside of these lines and gain prosperity and good fortune. If they were to locate directly on top of a line, they would purportedly become ill and disturbed.

Unfortunately, wide variations in technique and ability among those who know anything about ancient geobiological practices make it difficult to verify the locations of these lines today and consistently map them. Dowsing is the only tool available at the moment, and it has proven to be notoriously inaccurate. It is also possible that power lines, highways, and fence lines may have had some effect on the natural flows of invisible energy throughout the Earth as well. By using thought processes as shown in the Synergetics material, it may be possible to plot the probable location of natural ley lines, and then attempt to use observers, gravity wave detectors, scintillation counters, and magnetic detectors to pin down the exact current locations of these lines. This will go a long way towards easing the confusion that researchers are currently feeling in these matters.

TRANSPORTATION

Complete redesign of our transportation systems is essential if we are to prosper as a species. Many of the world's most serious pollution problems stem from the support of our current predominant method of transportation on this planet, which involves fossil fuel, and internal combustion or turbine engines. These engines create pollution, are expensive to manufacture, use tremendous amounts of natural resources, which then often become wasted, and are still slow, inefficient and dangerous.

According to some of the postulates of Synergetics, there is an abundance of natural energy existing as structural wave forms throughout the Universe. It is possible to find a way to hook into these energy patterns and use them for new and, as yet, undreamed of transportation methods.

Several of the legends that focus on the ancient Atlantean culture give another intriguing clue as to the nature of a possible transportation system. According to these legends, Atlanteans had built up a system of control crystals, which amplified certain aspects of the natural grid lines of the Earth, and allowed relatively simple electronic devices to tune into these lines of force and create a harmonic condition resulting in levitation of small vehicles. Whether or not these legends are true is open to great debate, but interestingly enough, Synergetics at least offers the possibility that such a transportation system could be developed. This is due to the fact that the structural, invisible lines of force that hold together any kind of astronomical body must contain some energy. There is no way around that. If one could tune into that energy somehow, and learn to resonate with it, then remarkable developments in this field would suddenly materialize.

By putting this idea into print, it is hoped that experimenters will delve into Fuller's methods of thinking, which should expand the frontiers of methodology and create these new transportation systems so desperately needed in our world today.

ENVIRONMENTAL REPAIR

The past five generations of humans on this planet have inflicted increasing amounts of environmental damage on landscapes and other species. Of course, this must stop and it will stop one way or another. It may stop with the sudden extinction of our species because of our planet's refusal to support us anymore. Or it may stop because we consciously choose to think in different ways, allowing us to stop doing any further damage and begin the difficult processes necessary to repair what we can.

We are already witnessing remarkable developments in chemistry and physics, such as Fullerene molecules, based on Synergetic forms. As we move deeper into these concepts, we will discover that all chemical elements are simply harmonic patterns held together by invisible structures which have an underlying consistency. The whole confusing system of quarks, muons, and mesons, and other strange transient subatomic particles, will no longer be necessary in attempting explanations of what is going on. Instead, we will be able to speak of elements and chemical compounds as harmonics resonant at certain frequencies. As we begin to understand the synergetic interactions between gravity, electromagnetism, g-force, and the harmonics of elements, we'll be able to learn how to transmute elements, and thus detoxify some of the worst pollution sites. Of course, this is tricky business and it is going to take a great leap in understanding from where we are.

METEOROLOGY

Storms track along spontaneously developed grid lines. Once we learn how to detect different kinds of grid lines that form and reform constantly, we'll be able to know in advance where a storm is going to track. These grids are simple in synergetic terms. Since there are invisible forces that pull the storms along, they should always be easy to map. We can then find the grid lines where storm tracks are anticipated several days in advance, and thus obtain warnings of floods, tornados, and hail.

It is now known, through the work of Trevor Constable, Peter Lindemann, Michael Theroux, and others, that forces associated with storms form in distinctive patterns. These patterns an be easily modified using simple radionic transmitters. In fact, the simplicity of effective devices is truly astounding. A small tube which seems to contain hardly any material at all can cause massive effects extending for miles in any direction. At this time, experiments with radionic weather devices are obviously dangerous. As we understand more, and begin to coordinate efforts along extensive communication lines, this knowledge can be collated and made useful.

CONCLUSION

Design Ecology is a new discipline, which has integrated a number of other disciplines. Each practitioner must make a personal choice regarding which parts are most important to pursue, depending on the needs of clients and one's own talents and abilities. Still, when not engaged in the study of any other matter, it would always be best for consultants to immerse ourselves in the Synergetics material. As we do that, we will improve our navigational abilities, enhance our creativity, and begin to understand a common mathematical language which describes the structure of our Universe. This understanding can easily be put to use for the benefit of our clients, as we grow in our own abilities.

THE FUTURE OF DESIGN ECOLOGY

One of the ways Design Ecology can help to improve conditions on this planet is to be a way to inform the design of new communities. Following is an article on one such community, which used every possible technique drawn from Bau-Biologie, solar design, and recycling water systems.

AUSTRIAN COHOUSING NEAR VIENNA

Vienna is said to be a city of miracles. Its not an easy town to live in it never was so it is a miracle for anyone who manages to survive there. For me, the miracle took a very strange form. While riding on the subway, my already strained back gave out. It was some of the most intense pain Id ever felt. Fortunately, the people who I was staying with knew the only acupuncturist practicing in Vienna, and they arranged an appointment. As everything turned out, my schedule for the next two days was completely blown because I was ordered not to ride a subway for at least that amount of time.

My hosts, knowing that I couldnt make a trip downtown, mentioned a friend of theirs who had some involvement with an intentional community west of the city. They asked if I'd like to talk with her. As it turned out, Freya Brandl, an architect and colleague of the developer, had a rare day off and had just purchased a new car which she was anxious to try out. So on a rainy February afternoon, we set out for Gartnerhof.

This is CoHousing at its best. Twenty families live in the community, which contains two connected sections. The front section, toward the road, has apartmentlike units. The rest of the community is groups of cluster houses, with two to four houses attached in each cluster. A large amount of open space extends out east of the housing units. All of the units are designed around pathways that cause residents to meet each other often, so the sense of community is heightened. However, the units themselves, although attached, are designed for privacy, with pleasant walls and courtyards.

Conservation was very much in mind while designing this community. Many of the units include solar greenrooms, which they call wintergardens. A few have supplemental solar hot water systems for heating.

The entire community shares an innovative water collection and recycling system. This system relies on collecting rainwater from underground cisterns, with a hookup to a municipal line as a backup. Since all the toilets are composting models, all that has to be treated is greywater used for washing and other household activities. This greywater is piped to three plant basins first, so the root systems of the plants accomplish initial water treatment. After that, the water runs through a series of flowforms, which look like concrete lilypads. These forms use a technology called vortexian mechanics, which is becoming popular in Europe. According to the principles of this technology as outlined by Viktor Schauberger in the 1930s, creating certain vortex shapes in flowing water helps to purify and energize the water. After the flowforms, the water settles into a lagoon, where it can be drawn back for recycling.

Solar Green Room at Gartnerhof

Helmut Deubner, the architect who developed this community and lives there, granted a short interview. He said the community took five years to plan before building. The entire design process was guided by the principles of BauBiologie, which means the biology of buildings. Everything, including the initial site plan, the building materials specified, and orientation of the units, was done with these principles in mind. The goal was to create living spaces that would optimally enhance health. Financing was through individual mortgage loans supplemented by some government grants for the innovative conservation technology. Each unit cost the equivalent of $118,000 in US dollars. Various government ministries are doing followup surveys on aspects of the community, because of their interest in the grants. Many of the residents have what Austrians consider alternative occupations, meaning craftspeople or independant consultants. Pottery made at the site is sold during the summer from one of the roadfront units. The community also has its own kindergarten.

In Austria, cohousing has some similarities with American projects. The most important differences are the potential for government involvement, and the presence of BauBiologie as a way to guide project design. Few projects exist at this time, so Gartnerhof is looked on as an inspiring model for the future by many Austrians.

House near Crestone, Colorado designed and photographed by architect Michael Bertin

CONCLUSION

The subject of this book will continue to evolve. New communities will spring up, which need to be documented. New techniques will be developed, and maybe even new sources of power to run our civilization. These all need to be communicated. The process of ephemeralization has already taken hold in many areas of the economy, especially in computers, housing, and now transportation. We need to mine less resources all the time, and are able to recycle more of what is used. Meanwhile, experiments go on daily in many fields, producing items using even fewer resources, and often generating less toxic wastes.

Each reader of this work is an essential part of the future of Design Ecology. Just because you have considered these ideas, and read this far, you've shown a committment on some level to making life decent for all of Earth's people. As you go out and implement these in your own way, you become a living experiment, and possibly a resource person who will help generations yet unconceived to live happy and full lives in this sparkling blue garden we call Earth.

I have a lot of hope for the future of our species. That's why this book was compiled, as a way to share ideas with other workers cooperating in this mission to improve our environment. If I didn't have hope, I'd be out there, as one of many faceless people at nameless parties, trying to blot it all out. For many years, that's how I actually lived. But the hope that inspired this book is usually with me now, and if it is passed on to you through this writing, that's even better. Thank you for participating in a better future for this planet.

RESOURCE GUIDE

EDUCATION

INSTITUTE FOR BAU-BIOLOGIE & ECOLOGY, INC.
Box 387
Clearwater, Florida 34615
(813)461-4371
Excellent correspondence course derived from original German materials. The course includes several packs of written material, and at least three seminars leading to certification as an Environmental Building Inspector. Shorter courses and seminars for the general public are also available.

INTERNATIONAL TESLA SOCIETY
EXTRAORDINARY SCIENCE CONFERENCE
P.O. Box 5636
Colorado Springs, Colorado 80931
(719) 475-0918
Source for information on advanced research into Geobiology, Free Energy, Resonance, and other relevant scientific topics. Hosts yearly symposia.

GEOMANCER'S BOOKSOURCE
Richard Feather Anderson
P.O. Box 1039
Sebastopol, California 94573
(707)829-8413
Excellent comprehensive catalog of resources, mainly relating to Feng Shui. Many of the books mentioned here are available through this catalog. Also provides classes and consultations.

BUCKMINSTER FULLER INSTITUTE
1743 S. La Cienega Blvd.
Los Angeles, California 90035
Provides a limited number of resources on Buckminster Fuller's work, and some educational tools such as modelling kits.

CONSULTANTS

THE ART OF PLACEMENT
Katherine Metz
1015 Gaylet Ave. #1218
Los Angeles, California 90024
310-208-5282
 General Feng Shui consulting, occasionally offers classes.

SEANN XENJA
(Formerly Thomas Howes Consulting)
Napa, California   94558
707-226-2248
SGXenja@aol.com
General Feng Shui consulting.

Elaine Jay Finster
P.O. Box 93
Bailey, Colorado 80421
(303)838-8446
Traditional Feng Shui, mostly according to principles of Professor
Lin Yun. Also expert on crystals and radionics.

APPRO TEC ENERGY MANAGEMENT
M. Spark Burmaster, E.E.
RR1, Box 77A
Chaseburg, WI 54621
Specialist in electromagnetic fields.

Slim Spurling
Golden, Colorado
(303)279-8324
Geobiology Consultant

Steven Maddox
Colorado Springs, Colorado
(719)576-2491
Specialist in lighting.

ENVIRONMENTAL TESTING & TECHNOLOGY
Peter Sierck
P.O. Box 230369
Encinitas, California 92023
Seminars & consulting services

HEALTHFUL HARDWARE
John C. Banta
P.O. Box 3217
Prescott, Arizona 86303
Seminars and catalog of equipment and supplies.

SAFE ENVIRONMENTS
David Bierman
2625 Alcatraz, #342
Berkeley, California 94705
Seminars & consulting services

PRODUCTS

REAL GOODS
P.O. Box 836
Hopland, California 95449
(707)744-2106
One of the best suppliers of photovoltaic (solar electric) items,
environmental products, housewares, composting toilets, more.

JADE MOUNTAIN
P.O. Box 4616
Boulder, Colorado 80306
(800)442-1972
Capsule flourescent bulbs, solar energy supplies, environmentally
conscious products.

CRYSTAL HILL FARM
Daniel Winter & Friends
9411 Sandrock Rd.
Eden, New York 14057
Catalog of materials relating to Sacred Geometry & Synergetics.

METAFORMS
Greg & Gail Hoag
P.O. Box 2262
Boulder, Colorado 80306
Wire sculptures using Synergetic principles, which can modify
subtle energies in a building.
MEDA
485 Spring Park Place #350
Herndon, VA   22070
(703)471-1445
Excellent gaussmeters.

HEALTHFUL HARDWARE CATALOG
P.O. Box 3217
Prescott, AZ   86302
(602)445-8225
Wide selection of meters, environmental products, and electrical
devices.

SEVENTH GENERATION
Colchester, Vermont 05446
(800)456-1139
Catalog including chemically untreated clothing & safe cleaning
supplies.

ALLERGY RESOURCES
P.O. Box 888
Palmer Lake, Colorado 80133
(800) 873-3529
Mail order catalog for EI patients

AN OUNCE OF PREVENTION
8200 E. Phillips Pl.
Englewood, Colorado 80112
(303) 770-8808
Mail order catalog of environmental materials

HAZCO
(800) 332-0435

This company provides an excellent catalog of environmental test equipment, personal protective equipment, and lab supplies, for sale or rent. This includes gas chromatographs, flame ionization detectors, and photoionization detectors. They have several branches in the USA.

PILLAR OF LIGHT ENTERPRISES
Jack Derby
2762 W. Desert Crest Drive
Tucson, Arizona 85713
Maker of the Violet Ray Crystal Resonator, an excellent device for well-being. Also manufactures a device which can produce a personal field which is resonant with the Earth.

TESTING LABORATORIES

P & K MICROBIOLOGY SERVICES, INC.
Chin S. Yang, PhD
3 Greentree Way
Cherry Hill, New Jersey   08003
Environmental testings services for professionals.

OZARK WATER SERVICE AND LABORATORY
P.O. Box 218
Sulphur Springs, Arkansas 72768
Provides do-it-yourself water test kits and water purifiers.

SUPPLEMENTAL NOTES ON SOURCES

INTERVIEWS WITH EI PATIENTS

Almost everyone interviewed in connection with this section wished to remain anonymous. Unfortunately, EI carries a social stigma in many circles, and reports of persecution against patients by medical professionals, insurance companies, and even family members do circulate. Many patients have banded together in formal and informal associations. One of the more accessible organizations in Colorado is the Rocky Mountain Environmental

Health Association. Several of their members were helpful. Listed here are a few of the most important interviewees by pseudonymous initials only:

T.C.: Was sensitized through intense exposure to cigarette smoke while working in a chemical manufacturer's office in the mid-1970's. Several co-workers have filed suit against the manufacturer because of this exposure. Is now sensitive to many chemicals, especially scented deodorants and Clorox bleach. Uses alternative cleaning products in her home, which she says are not as effective as chemical products. Has had problems with family members, who at one point tried to have her involuntarily committed to a mental institution. Discovered through experience that no conventional therapy worked to control her sensitivities, although alternative therapies met with some success.

C.T.: Accidentally exposed to a now-banned pesticide, EDB, while on a Forest Service work crew in the 1970's. In 1992, attempted to move back into Denver after having been alone in the mountains for several years. Her state of recovery only allowed her to stay about six months in the city, and she ended up moving back to the cabin.

W.W.: Exposed to Phenol in her youth, from mosquito spraying trucks. Has chronic mental impairment, but has learned to adapt and compensate.

R.C.: Must live in a wheelchair because of exposure to pesticide sprays from an anti-marijuana campaign. The most difficult patient to interview -- has extreme problems with even the simplest of daily tasks and also suffers from MCS. Her prognosis is not good, and she may not live much longer.

G.R.: Employee at the special industrial pre-treatment water plant for radioactive wastes in Grand Junction.

G.G.: Developed MCS over a period of several years due to exposure to chemicals in his workplace. Demonstrates mild sensitivity to printer's inks and intense sensitivity to perfumes. One day while we had lunch together, we had to move because a woman at the next table was wearing a strong perfume. Currently works doing environmental air quality surveys; says he's "an ideal canary" because of his chemical sensitivities.

HELPFUL THERAPISTS

Unfortunately, the few therapists willing to deal with EI and MCS cases are so swamped with work it is almost impossible to line up interviews. One therapist in the Denver area, who has a national reputation, has a six-month waiting list for appointments. Compounding this problem is that fact that some medical boards and insurance companies are openly hostile to people dealing with these problems. A few doctors who have attempted new treatment methods for EI have been outrageously persecuted by their professional boards.

Due to fear of reprisal from the American Medical Association and civil authorities acting under AMA orders or in sympathy with AMA goals, most of the therapists consulted for the Treatment Protocols section requested that their names not be printed. Therefore, as a professional courtesy, none of the names are being used. Three of the therapists consulted have actually been jailed for treating patients, even though their treatments were effective. The total number of therapists consulted to develop this section exceeds ten.

Interviews conducted in February, 1992 with a physician practicing in Detmold, Germany, who is sympathetic to homeopathy, and with members of a group associated with a manufacturer of homeopathic preparations based in Switzerland contributed to the Treatment Protocols section. Notes and supportive material from the European trip are available to my clients by request.

MISCELLANEOUS NOTES

Relevant explanations of material from Selected Writings of Wilhelm Reich were provided by Klark Kent, of Dayton, Ohio.

John Ott, who got his start researching the technical aspects of light for Walt Disney Productions, is generally recognized as the world's foremost expert on this subject. If you want more information, his magazine articles are an excellent source.

REVISED METEOROLOGICAL NOMENCLATURE:
A version of this paper was originally published in: Tesla 89, Vol.5 Number 4 Oct/Nov/Dec 1990

ANNOTATED REFERENCE BIBLIOGRAPHY

GENERAL REFERENCE

Correspondence Course Baubiologie
Dr. Anton Schneider, Helmut Ziehe translator
Institute for Bau-Biologie and Ecology 1988
There is nothing more comprehensive available anywhere.

FENG SHUI
Presented in order of importance for understanding basic concepts.

INTERIOR DESIGN WITH FENG SHUI
Sarah Rossbach
Dutton, 1987
A good, practical text on this subject. Photographs and diagrams are very helpful. This is perhaps the most well-known book on Feng Shui, drawing heavily from Professor Lin Yun's Black Hat techniques.

THE BOOK OF CHANGES AND THE UNCHANGING TRUTH
Ni, Hua Ching
1983
Besides being an excellent rendition of the I CHING into English by a native Chinese speaker, this also contains a large amount of cultural context material. By using this along with the Wilhelm/ Baynes translation, you can decode many of the more subtle aspects of Chinese physics, and obtain a good working

understanding of how to gain experience in Feng Shui.

HEALTH, WEALTH & BALANCE THROUGH FENG SHUI
Elaine Jay Finster
P.O. Box 93
Bailey, Colorado 80421
This book contains a good summary of several key teachings of Lin Yun, one of the most accessible Feng Shui experts in America.

FENG SHUI -- A LAYMAN'S GUIDE TO CHINESE GEOMANCY
Evelyn Lip
Heian, 1987
Filled with diagrams and photos, this is a very good introductory work. Especially recommended for people who want to know about Feng Shui but don't have a lot of time to study.

THE LIVING EARTH MANUAL OF FENG-SHUI
Stephen Skinner
Arkana, 1989
To accomplish deeper study, this is a good book to have. It brings out the mathematical basis of the "five elements" system, and imparts practical instructions on using a Chinese compass.

TAO TE CHING
Gia-Fu Feng & Jane English
Vintage Press, 1972
This is the most widely used current translation of the work. In the 1960's, several other translations were available, including an especially good one published by Penguin. It's good to compare different translations. It's best to learn Chinese, if you want to have a better chance of understanding this work.

CHUANG TZU
Burton Watson
Columbia Univ., 1968
While not as well known as the Tao Te Ching, this book gives several key concepts which are common to both Taoism as a religion and Feng Shui as a science. It is especially good for perceptual re-training.

I CHING
Wilhelm/Baynes
Princeton
Most serious students obtain this translation. It is scholarly, very credible, and also respectful of the people who have lived with it for many generations. However, it is fairly difficult to read, mainly because of the way it is arranged.

THE I CHING AND ITS ASSOCIATIONS
Diana ffarington Hook
Penguin, 1980
Shows in precise terms how the I Ching can be used for the study of Feng Shui patterns. Difficult to find in the USA.

BOOK OF THE HOPI
Frank Waters
Ballantine 1963

CHEMICALS AND ENVIRONMENTAL ILLNESS

THE E.I. SYNDROME
Sherry A. Rogers, MD
Prestige 1986
Since the writer is a physician, this book has unusual credibility. In Rogers' case, she suffered from Environmental Illness for many years, and several of the treatment protocols mentioned in the book were necessary for her own survival. Its poor organization makes it difficult to use as either a reference or reading material, but with persistence a great deal of valuable information can be found in here, as recorded by a courageous pioneer.

CLINICAL TOXICOLOGY OF COMMERCIAL PRODUCTS
Gosselin, Smith, Hodge
Williams & Wilkins 1984
Medical personnel often use this basic reference in developing diagnoses and treatments. It was consulted extensively for the sections on individual chemicals and chemical families.

THE COMMON SENSE APPROACH TO HAZARDOUS MATERIALS
Frank L. Fire
Fire 1986
Intended as an emergency responder's course textbook, this was open throughout much of the writing of this section. It provides one of the best overviews of this field available today. The focus is on handling emergencies and acute toxicity, often from a fireman's perspective.

A CONSUMER'S DICTIONARY OF HOUSEHOLD, YARD, AND OFFICE CHEMICALS
Ruth Winter
Crown 1992
This is a reference for every conscious consumer. Winter tends to avoid technical terminology, in order to make the work more useful to larger numbers of people. Since the book is arranged as a dictionary, it is easy to find what you're looking for.

A BREATH OF FRESH AIR
LaRonna DeBraak
Mountain Meadow Publishing
P.O. Box 21353
Denver, CO 80221
A practical guide for filtering out indoor air pollution by using houseplants.

THE NONTOXIC HOME
NONTOXIC AND NATURAL
Debra Lynn Dadd
Tarcher
These books are written from a layperson's point of view, and thus contain a lot of valuable information about household chemicals in an easily readable form. In her more recent book, NONTOXIC AND NATURAL, much of the material is updates and many references to alternative product suppliers are listed.

NIOSH POCKET GUIDE TO CHEMICAL HAZARDS
National Institute for Occupational Safety and Health 1990

178

U.S. Government Publication
It is impossible to cover every hazardous chemical in use today in one volume, but this booklet is a good start. It is regarded as a good basic reference by many workers in industrial hygiene, hazardous material handling, emergency response, and similar fields.

POISONING
Jay Arena, M.D.
Thomas 1979
Many classes of industrial and household chemicals are covered in an easily accessible format. This is a good general reference on toxicology, and was especially helpful for the sections on Formaldehyde and the Aromatics.

NUKLIDKARTE
several authors
Kernforschungszentrum Karlsruhe 1981
Every known isotope of every chemical element is listed here. It was extensively consulted for the section on Radioactivity. Unfortunately, it does not seem to be available in the United States. I found my copy in Munich.

A CITIZEN'S GUIDE TO RADON
EPA 1986
RADON REDUCTION METHODS
EPA 1989
U.S. Government Publications
Two pamphlets often found together on information tables. They offer some of the most sensible information available on this topic, and can help prevent contracting fraud, which is now rampant in several regions of the United States, as unscrupulous operators play on public fears.

CHRIS HAZARDOUS CHEMICALS DATABASE
United States Coast Guard
These are data sheets contained in three binders, available at many public and trade school libraries. Much of the information is oriented toward water pollution control and acute toxicity to

emergency responders. Basic TLV, Odor Threshold, and IDLH values were generally drawn from these.

THE HERBALIST
David Hoffman
CD-ROM available through:
HOPKINS TECHNOLOGY
421 Hazel Ln.
Hopkins, Minnesota 55343
(612)931-9376
Everything you would ever want to know about herbal medicine is contained in an easily approachable format.

ELECTROMAGNETISM

Papers by Alexis Guy Obolonsky and Col. Thomas Bearden are contained in both the 1986 and 1988 Proceedings of the International Tesla Symposium. Contact the International Tesla Society at P.O. Box 5636, Colorado Springs, Colorado 80931

THE ION EFFECT
Fred Soyka & Alan Edmonds
Dutton 1977

ELECTROMAGNETIC POLLUTION SOLUTIONS
Dr. Glen Swartout
Available through INSTITUTE FOR BAU-BIOLOGIE & ECOLOGY, INC.

GLOSSARY

BAU-BIOLOGIE

In Germany, this term is commonly understood. It literally means "the biology of buildings". It is a method of considering and evaluating all environmental factors present in a living or working area.

CH'I

According to most ancient Chinese writings, this is the fundamental energy of the Universe. Water is often used as a symbol of this energy. It cannot be seen or measured. Modern insights indicate this energy is closely related to sound waves.

CLINICAL ECOLOGY

Medical method of considering all possible factors in an environment when diagnosing an illness. This is controversial within the medical community.

DESIGN ECOLOGY

A coined term, made up as a correlation to the medical term "Clinical Ecology", which is a method of looking at the entire body in conjunction with environmental factors. The entire environment is assessed in relation to what can affect body, mind, and spirit.

DX

Means "distance" as applied to radio signals. It is an old amatuer radio operator abbreviation, dating back to the days when only Morse Code could be transmitted and everything was abbreviated.

ELECTRO-STRESS

A test run sometimes on individuals. It measures the amount of voltage actually detectable in the human body. The higher the voltage in the body, the more likely it is that stress-related health problems would develop.

ENVIRONMENTAL ASSESSMENT

Any checking of what may be in an environment. This can be focused or more broad. Mostly, this term is associated with governmental actions and requirements, and so can be confusing when applied to private practice.

ENVIRONMENTAL AUDITING

In the late 1970's and early 1980's, when tax breaks for alternative energy installations were available to American homeowners, "energy audits" were commonly performed. These were usually done by solar system salesmen or insulation contractors, although there were a few independents who specialized in this. An energy audit was a procedure using infrared thermometers, smoke tests, and careful observation to determine points where energy leaks out of a building. Then, insulation and solar energy supplementation would be recommended where appropriate. This procedure was largely forgotten after the tax breaks were discontinued in 1985.

Survey techniques derived from Bau-Biologie and adapted to American needs could be called an "environmental audit".

FENG SHUI

A Chinese term for environmental consulting services, which dates back about 6,000 years. It can be called "acupuncture for buildings", since it embodies the same set of principles which are the foundation of most acupuncture practice.

FLAME IONIZATION DETECTOR (FID)

A piece of equipment which detects the presence of chemicals by burning small amounts of air and automatically analyzing the reaction. It cannot be used in flammable environments.

FLUORESCENT LIGHTING

Lighting obtained by running an electrical charge through a lamp or tube coated with a substance which then glows.

GEOBIOLOGY

Study of invisible energies associated with the Earth, as

they affect living beings. These energy patterns fall into many classes. This is a new field, and ideas about what is happening are always subject to modification.

GEODESIC

Pertaining to geometric lines which are edges of regular shapes derived from spheres. A geodesic dome makes use of these lines to create an efficient, orderly shape.

HVAC

Abbreviation for Heating, Ventilation, and Air Conditioning.

INCANDESCENT

Refers to lighting generated by heating an element, as in ordinary light bulbs.

ION

Any atom which has an unbalanced electrical charge because of too few or too many electrons. Fully ionized atoms have no electrons present.

METABOLITES

Secondary substances formed in the body, which are combinations of toxic materials and natural body chemicals. Sometimes these are more poisonous than the original pollutant.

MICROWAVES

Electromagnetic radiations at extremely high frequencies, typically vibrating over a billion times a second. They can only be transmitted in a line of sight, and will be stopped by metal or rock. They are used for communications, satellite operations, and cooking.

MITIGATION

The process of trying to neutralize an environmental pollutant. This can be done by stabilizing it in place, altering it to something harmless, or removing it from the area.

MONOMER
A molecule of an organic compound which can combine with similar molecules to form long chains, called polymers. Often, this is an intermediate material in plastic manufacturing.

NIOSH
National Institute for Occupational Safety and Health, a research organization funded by the United States government. It is charged with gathering the data used to make safety regulations.

OSHA
Occupational Safety and Health Administration, agency of the United States government which is supposed to make sure all workplaces are safe for workers. It has been well-known for drafting unenforceable regulations while sometimes ignoring actual threats to workers, and so is often used as an example of government ineffectiveness.

ORGANIC
In chemistry, this refers to any compound which contains carbon. In commerce, this is supposed to mean foods and fibers grown and processed without pesticides and chemical fertilizers.

ORGONE
An energy form first discovered by Dr. Wilhelm Reich, during his research work covering the 1930's to the 1950's. He invented several devices which would concentrate this energy, and also found that it is highly toxic in the presence of radioactive material. It seems to be somehow related to "scalar waves", "zero point energy", and possibly "radionic rays".

OUTGAS
Vapors of a substance may slowly evaporate into the surrounding air. This is a problem with certain building materials.

PHOTO-IONIZATION DETECTOR (PID)
A piece of equipment which detects the presence of chemicals through a strong light beam. It is not as accurate as a Flame Ionization Detector, but can be used in flammable areas.

POLYMER

A long chain-type molecule which can be exceptionally strong. Many plastics are polymers. In manufacturing these, other substances are used which may cling to the polymer molecules, and which may also be toxic.

QI

Another way to spell Ch'i in English, as officially mandated by the People's Republic of China.

RADIONICS

Study of invisible radiations which are not classified by contemporary physics. Results of this research include devices which can generate healing effects at a distance. In the USA, many radionic devices have been confiscated by the government.

SCALAR INTERFEROMETER EFFECT

Whenever anything has two poles, it will naturally tend to send and receive wave forms. These waveforms generally are of the type related to Ch'i energy, or other phenomena not classified by physics.

SCALAR

Pertaining to magnitude, but with no direction in space, therefore non-locational. Can be applied to volume or temperature, as well as other forces.

Many researchers have been attempting to extend this definition in various ways, applying it to forces in the universe which we do not yet understand, and there is presently a large amount of disagreement about the exact definition as applied to advanced scientific research.

SENSITIZER

Any chemical which has caused a person to become sensitive to other chemicals.

TACHYON

Originally, this was a hypothetical fundamental particle in physics, characterized by fast movement. It has never been found.

Since the 1980's, the term has been applied by several researchers and commercial groups to unspecified waveforms or particles which are emanated by their products.

VOLTS
This technical term is a measurement of the amount of "push" happening in an electrical circuit.

ELECTROMAGNETIC MEASUREMENT TERMINOLOGY

TERM		SAFE LEVEL BELOW:	DEFINITION
mG	= milliGauss	2.5	Magnetic fields
mV/m	= milliVolts per meter in the air	20	Electric fields, actual charge
mVDC	= milliVolts DC	200	translated comparative reading of High Frequency & Microwave fields
VAC	= Alternating Current voltage (Electro-Stress)	1.1	
kV	= kiloVolts	1,000	Volts (in powerlines)
mR/hr	= milliRems per hour	0.3	comparative measurement of ionizing radiation

Index